Zero Emissions Power Cycles

Zero Emissions Power Cycles

E. Yantovsky

J. Górski

M. Shokotov

CRC Press
Taylor & Francis Group
Boca Raton   London   New York

CRC Press is an imprint of the
Taylor & Francis Group, an **informa** business

CRC Press
Taylor & Francis Group
6000 Broken Sound Parkway NW, Suite 300
Boca Raton, FL 33487-2742

Library of Congress Cataloging-in-Publication Data

Yantovsky, Evgeny.
Zero emissions power cycles / Evgeny Yantovsky, J. Górski, and M. Shokotov.
p. cm.
"A CRC title."
Includes bibliographical references and index.
ISBN 978-1-4200-8791-8 (hardcover : alk. paper)
1. Fossil fuel power plants--Equipment and supplies.. 2. Fossil fuel power plants--Energy conservation. 3. Pollution control equipment. 4. Internal combustion engines. 5. Electric power production--Environmental aspects. I. Górski, J. II. Shokotov, M. III. Title.

TK1041.Y36 2009
621.31'2132--dc22   2008042182

Visit the Taylor & Francis Web site at
http://www.taylorandfrancis.com

and the CRC Press Web site at
http://www.crcpress.com

Contents

Preface

Biographical Notes

Acronyms

Preface

The idea for this book originated long ago in discussions with Professors David A. Frank-Kamenetsky, John O'M. Bockris, and Gustav Lorentzen.

The present text is compiled from the papers published since 1991, alone or with colleagues including Professors J. A. McGovern, J. W. Gorski, M. K. Shokotov, and N. N. Akinfiev. I am very thankful to all of them for the permission to integrate and use these works. The papers are listed in the references.

At the beginning of the twenty-first century, it is possible to speak realistically about electric power production burning fossil fuel without pollution, by using developments in materials, combustion technologies, cooling techniques, and creative cycle concepts. Alternative approaches to the production of power using fossil fuels without pollution have been proposed in the United States and Europe, and these proposals are being continually improved. This volume describes and evaluates some existing alternative approaches for the potential benefit of students, industry members, and regulators. This work does not deal in detail with the so-called renewable technologies, such as photovoltaics (solar cells), wind turbines, tidal motion systems, or geothermal systems, because these systems do not employ fossil fuel for combustion as a means of generating electricity. In this work, the focus is on fossil-fuelled, nonpolluting power generation systems. Numerous alternative technologies are emerging and we seek here to explain these technologies and their relative merits. The topics and technologies described in this work are continually advancing, but the basic concepts and operational characteristics of the systems described are not expected to change significantly in the coming years. We believe the time is right to record and describe some of the new emerging technologies. The G-8 meeting of world leaders set a goal of "50% till 2050" to reduce emissions (with a multi-trillion dollar funding). It makes for great business with the building of hundreds of zero emissions power plants.

Note: in the book, Chapters 3 and 6 have been written with Jan Gorski, and Chapter 7 and Section 9.3 were written with Mykola Shokotov.

I wish to thank leading researchers of Clean Energy Systems, Inc., Dr. R. Anderson, and Dr. S. Doyle for their help in editing the first two chapters of the book and for the moral support of the book publishers.

E. Yantovsky
Aachen, Germany

Biographical Notes

E. Yantovsky, Ph.D., was born in Kharkov, Ukraine, in 1929. His main areas of interest are magnetohydrodynamic generators and pumps, heat pumps, zero emission power plants with membrane oxygen for combustion, energy and exergy currents, and exergonomics. Professor Yantovsky graduated from the Aviation Institute in Kharkhov and moved to Taganrog to work on the production of large hydroplanes. In 1953, he returned to Kharkov and worked in a large electrotechnical plant where he was responsible for testing the air and heat flow of electrical machinery. For a short time, he worked as a designer of large synchronous motors. From 1959 to 1971, he was the head of the MHD Laboratory in Kharkov, where the magnetohydrodynamic (MHD) liquid metal generator was built and tested, with the intention of powering a space ship to Mars. professor Yantovsky joined the Krjijanovski Energy Institute in 1971 as a senior researcher, and then in 1974, he began work in the same capacity for the Institute for Industrial Energetics. Between 1986 and 1995, Professor Yantovsky was chief researcher at the Energy Research Institute of the Russian Academy of Science. He has also undertaken numerous visiting lectureships throughout Europe and the U.S. As an author, Professor Yantovsky has written six books and published over 40 articles in English (including *The Thermodynamics of Fuel-Fired Power Plants without Exhaust Gases* in 1991 and *The Concept of Renewable Methane* in 2000). He lives in Aachen, Germany.

(From *Founders of 21st Century,* Intern. Biogr. Centre, Cambridge, 2003, p. 661).

Jan Gorski, Ph.D., D.Sc., was born in 1945 in Letownia, Poland. He is a specialist in the area of applied thermal sciences and energy conversion systems. The particular subjects of his research interest are the problems of dense gas phenomena in the thermodynamic and flow process simulation. In over 30 years of professional experience, he has worked in the aeronautical industry as a gas turbine design engineer. Since 1974, he has been a tutor and an associate professor at both the Mechanical & Aeronautical Faculty of the Rzeszow University of Technology and the Faculty of Civil & Environmental Engineering. He is a member of EUROMECH and a member of two committees of the Polish Academy of Sciences.

Dr. Gorski has also worked as a visiting professor at the Universidad Nacional Autonóma de México in 1982 and has given a number of lectures throughout the European Union.

Mykola Shokotov was born in 1926 near the city of Lugansk, Ukraine. Between 1943 and 1945, he served in the military in WWII, including the storm of Berlin. Between 1950 and 1955, he attended the Kharkov Polytechnic Institute (KhPI), followed by three years as a designer of turbines and superchargers in a piston-engine factory. For 40 years, he worked as a lecturer then docent and professor in the Department of Internal Combustion Engines at KhPI. Along with lecturing, he was engaged in the scientific development of diesel engines for transport and industry. For many decades, he was a consultant to a
large Kharkov mill, manufacturing diesel engines for the transport sector. Professor Shokotov has published about 200 papers in technical journals and six textbooks. Since 1998, he has lived in Germany.

Acronyms

AZEP	Advanced zero emission power
BFW	Boiler feed water
C	Compressor
CAR	Ceramic autothermal recovery
CC	Combustion chamber
CES	Clean Energy Systems, Inc.
CHP	Combined heat and power
COOPERATE	CO_2 prevented emission recuperative advanced turbine energy
CW	Cooling water
DOE	Department of Energy
EG	Electric generator
EM	Electric motor
EOR	Enhanced oil recovery
ECBM	Enhanced coal bed methane recovery
ESA	European Space Agency
FBC	Fluidized bed combustor
FT	Fuel tank
GHG	Green house gas
HE	Heat exchanger
HP	High pressure
HPT	High pressure turbine
HTT	High temperature turbine
HRSG	Heat recovery steam generator
HX	Heat exchanger
IEA	International Energy Agency
INJ	Fuel injector
IP	Intermediate pressure
IPT	Intermediate pressure turbine
IGCC-MATIANT	Integrated gasification coal cycle MATIANT
ITM	Ion transport membrane
ITMR	Ion transport membrane reactor
L	Luft (air)
LHV	Lower heating value
LP	Low pressure
LPT	Low pressure turbine
MATIANT	Cycle designed by MAThieu and IANTovski
MCM	Mixed conducting membrane
MHX	Multi-heat-exchanger
MPT	Mean pressure turbine
NG	Natural gas

OITM	Oxygen ion transport membrane
P	Pump
PE	Piston engine
R	Radiator
RH	Reheater
SOFC	Solid oxide fuel cell
TC	Air turbocompressor
TIT	Turbine inlet temperature
WS	Water separator
ZECA	Zero Emission Coal Alliance
ZEMPES	Zero emissions membrane piston engine system
ZENG	Zero Emission Norwegian Gas
ZEITMOP	Zero emissions ion transport membrane oxygen power
ZEPP	Zero emission power plant

1 Controversial Future

1.1 INTRODUCTION AND FORECAST

Energy policy is full of controversies. Some people believe that fossil fuels will soon run out and contend that we need to think more about nuclear and renewable energy sources (solar, wind, geothermal, tidal) especially in view of perceived global warming. Greenhouse warming is caused by the greenhouse effect of the Earth's atmosphere. Some contend that warming is caused by the emission of combustion products from power plants and motor vehicles. Other people, including the authors of this book, believe that enough fossil fuels (gas and coal) are available to last at least for this century, that nuclear power is dangerous, and the German government's decision to eventually shut down all the German nuclear reactors may be a precedent for other countries. Renewables are beneficial alternative energy sources but only for the next century. Because of their high cost, renewables cannot dominate the energy supply for this century.

The world energy balance for 2030, as forecast by the International Energy Agency (IEA) and shown in Figure 1.1, illustrates the expected trends in energy supplies.

The IEA forecast anticipates the nuclear share to significantly decrease, while the share for renewables remains the same, and the share for fossil fuels is projected to increase by 2%, from 80% in 2002 to 82% in 2030.

The projected increase in the percentage of energy being provided by hydrocarbon fuels is important in view of the irreconcilable controversy between "peakists" and "non-peakists." The former are certain that the world soon will run out of oil and some years later, run out of gas, which will lead to serious economic problems, possibly escalating to wars over oil. The crucial point is that there will be a peak of oil production with a gradual decline afterward. Graphs of production versus time are depicted by a bell-shaped curve.

Specialized groups of "peakists" actively promote their views in the public media and on numerous Internet sites (e.g., www.energy-bulletin.net, and www.lifeafter-theoilcrash.net). Their opponents, "non-peakists," contend that a peak in oil production due to dwindling resources has been forecasted many times, and all of the forecasts were wrong. The oil resources are finite in principle but still rather large. Natural gas, mainly methane, is chemically much simpler than oil and, according to the abiogenic theory of gas origination, it might be produced naturally in the Earth's depths. The discovery of seas of liquid methane on cold planets adds credence to that theory. The most radical "non-peakists" refer to the "peakists" as "professional pessimists."

The world's largest oil company, Exxon Mobil, concluded on their website (www.exxonmobil.com) that, "With abundant oil resources still available — and industry, governments and consumers doing their share — peak production is nowhere in sight."

Considering the forecast of the IEA, the most authoritative international body in the energy field, the authors believe hydrocarbon fuels, including coal, will be

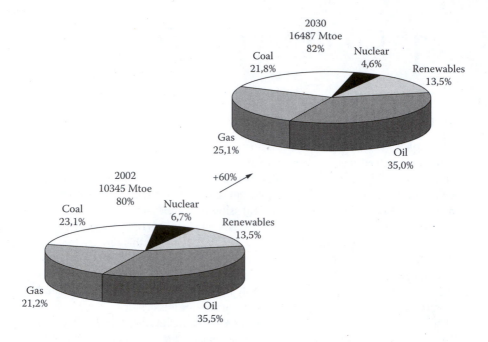

FIGURE 1.1 Forecast of the International Energy Agency published in 2004 on the world energy balance in 2030. (Data from World Energy Outlook, 2004.)

the dominant fuels in the 21st century. This means the primary impediment to their continued use is atmospheric pollution, which is a real cause for concern. This book shows alternative ways to prevent such pollution.

The reality of global warming now seems to be beyond controversy. It definitely exists. Not only precise measurements but also anecdotal evidence, such as the melting of polar ice and other climatic events, reveal the global temperature is increasing. It is now clear that there is a problem. Controversy remains over the appropriate response and the solution to the problem. This controversy generates two questions: (1) who is responsible, and (2) what should be done?

The most authoritative evaluation to date of the first question has been done by the United Nations' Intergovernmental Panel of Climate Change (IPCC), organized jointly by the World Meteorological Organization (WMO) and the United Nations Environment Program (UNEP). In its most recent report, the IPCC stated that climatic changes seen around the world are "very likely" to have a human contribution. By "very likely," the IPCC means greater than 90% probability.

At the launch of its Fourth Assessment reports, the IPCC chairman, Dr. Rajendra Pachauri stated the following: "If one considers the extent to which human activities are influencing the climate system, the options for mitigating greenhouse gas emissions appear in a different light, because one can see what the costs of inaction are."

The accuracy of IPCC forecasts is evident. The IPCC 2001 report forecast a temperature increase in 5 years as 0.15 to 0.35°C. The actual measured change was 0.33°C,

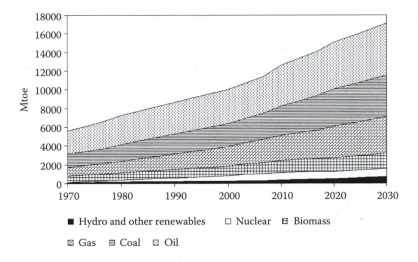

FIGURE 1.2 World primary energy demand by fuel in the reference scenario. (*Source*: World Energy Outlook 2006, International Energy Agency, OECD/IEA, 2006.)

very near to the top of IPCC range. The world energy demand as forecast in the 2006 IEA report and shown in Figure 1.2 offers the same forecast as in Figure 1.1.

The authors believe that the domination of fossil fuels as an energy source is beyond dispute and will persist for decades. This presents a real problem considering the gradually increasing share of energy that oil and gas are anticipated to provide.

Historically, world generation of electrical energy and the forecast through 2030 are even more significant. The history and forecast are presented in Figure 1.3.

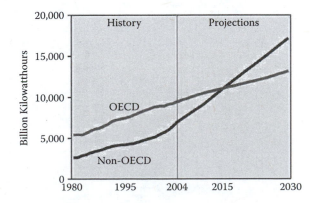

FIGURE 1.3 Historic and projected world electrical power generation by region. (*Source*: History: *International Energy Annual* 2004, Energy Information Administration (EIA), May–July 2006, (www.eia.doe.gov/iea); (Projections: System for the Analysis of Global Energy Markets, Energy Information Administration (EIA), 2007.)

From Figure 1.3, it is evident that significant growth is expected in the Organisation for Economic Co-operation and Development (OECD) sector and the growth in the non–OECD sector is even more pronounced. China and India play major roles in the latter sector.

1.2 REASONS FOR CLIMATE CHANGE

The possibility of global warming due to increasing concentrations of carbon dioxide in the atmosphere was forecast by Joseph Fourier in 1827 and, in more detail, by Svante Arrhenius in 1896. M. I. Budyko, in 1970, gave a description of the phenomenon, using all available geophysical information. Many controversies have appeared, pointing out not only warming, but also cooling effects in the atmosphere. Recent data have confirmed the excess of warming as measured in radiative forcing (i.e., positive and negative energy currents expressed in watts per square meter) and the skyrocketing increases in the concentration of greenhouse gases (CO_2, ppm, CH_4, ppb, and N_2O, ppb) in the atmosphere which are major forcing factors. The radiative forcing currents and the concentrations of greenhouse gases are shown in Figure 1.4.

"Positive forcing" means warming, whereas "negative forcing" represents cooling. From Figure 1.4, it can be seen that in ancient time, before human impact, positive and negative forces were almost equal with the possibility of cooling (ice ages). In recent times, the concentrations of greenhouse gases have skyrocketed, producing strong positive radiative forcing.

The authors thus conclude that there is incontestable evidence that global warming is occurring. Qualified international cadres of experts also believe that there is a significant human contribution to the problem, that it is essential for humanity to address this issue, and reasonable and effective ways to minimize human contributions to global warming must be found.

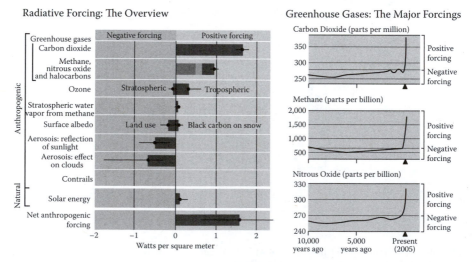

FIGURE 1.4 Overview of radiative forcing factors and the concentration of greenhouse gases in the atmosphere (*Sci. Am.* 8/2007, 67. With permission of the artist, Deaniela Naomi Molnar).

1.3 CONTROVERSIAL STATEMENTS

In an e-mailed statement, Claude Mandil, the executive director of the Paris-based agency, which advises 26 consuming nations, stated, "Under-investment in new energy supply is a real risk. The energy future we are facing today is dirty, insecure, and expensive." (Nov. 7, 2006).

The IEA stated in its Annual World Energy Outlook that governments and companies are expected to spend $20 trillion on power generation in the next 25 years, and there is "no guarantee" it will succeed. More than half of the total will be needed for emerging countries, where both demand and supply are rising the fastest. It is estimated that China alone will need to spend $3.7 trillion on energy between 2005 and 2030.

Forty years ago, the combustion of organic fuels without emissions into the atmosphere was proposed. In 1967, Degtiarev and Grabovski (see Figure 2.2) published schematics showing a method to capture carbon dioxide in an ordinary power unit for its subsequent use as a raw material in industry. In 1977, Marchetti proposed capturing carbon dioxide from fuel-fired power plants and sequestering it deep in the ocean or underground, thus protecting the atmosphere (see Marchetti, 1979).

Since these early proposals, the idea has been gradually expanded as well as demonstrated. The most recent indicative event was the 6th Annual Conference of Carbon Capture and Sequestration held in May 2007, in Pittsburgh, PA, U.S.A., during which many real and developing Carbon Capture and Sequestration (CCS) projects were explored.

To evaluate the scale of anticipated business growth, a recent report by the Massachusetts Institute of Technology (MIT) mentioned in *Power Engineering International* No. 5 (2007):

> ...CO_2 capture and sequestration (CCS) is the critical enabling technology that would reduce carbon emissions significantly, while also allowing coal to meet the world's pressing energy needs.... With CCS more coal is used in 2050 than today... A major contribution to global emissions reduction for 2050 is the reduction in CO_2 emissions from coal to half or less of today's level.

The MIT report calls for a successful large-scale demonstration of the technical, economic, and environmental performance of the technologies that make up all the major components of an integrated CCS system.

The figures forecast by MIT for annual coal use (in $EJ = 10^{18}J$) and CO_2 emissions (in $Gt = 10^9$ ton) are shown in Table 1.1.

Assuming that the forecast coal use with expanded nuclear and CCS in the U.S.A. in 2050 is used for power generation only, the 25 EJ of energy would produce 1000 GW of thermal power or about 500 GW of electrical power. This means that in 40 years, approximately 500 big coal-fired plants (1000 MW_e each) or about 12 per year, would need to be built in the U.S.A. Assuming coal-fired power plant capital costs to be approximately $2000 per kW_e, the total required investment is about $1000 billion. To accomplish such a multibillion dollar business, accumulation of investments for this purpose need to be amassed immediately.

TABLE 1.1

MIT Forecast of Annual Coal Use and CO_2 Emissions Assuming Several Different Energy Development Paths

| | | Business as Usual | | Limited Nuclear | | Expanded Nuclear | |
		2000	2050	With CCS	Without	With CCS	Without
Coal Use	Global	100	448	161	116	121	178
	U.S.A.	24	58	40	28	25	13
	China	27	88	39	24	31	17
CO_2 Emissions	Global	24	62	28	32	26	29
	From Coal	9	32	5	9	3	6

Source: Katzer, J. et al., 2007.

Unfortunately, CCS technology is still ignored at high political levels in some places. The Environmental Committee of the European Union, in its 3rd Report on Climate Change, was against new coal-fired power plants in Germany. This might affect the policy of "Atomausstieg," meaning to step away from atomic power. The best course for a coal-rich Germany would appear to be the earliest possible implementation of coal-fired CCS (zero emissions) power plants. The results of testing a 30-MW_{th} pilot-scale zero emission power plant in 2008 at Schwarze Pumpe by Vattenfall will be critically important as a demonstration of coal-fired zero emission technology in Germany.

In general, the policy of the European Union seems to be appropriate, as seen in the organization of a Technology Platform FP7 with the statement:

> In line with the proposed priority for "Near Zero Emission Power Generation" in FP7, the initial scope of the Platform aims at identifying and removing the obstacles to the creation of highly efficient power plants with near-zero emissions, which will drastically reduce the environmental impact of fossil fuel use, particularly coal. This will include CO_2 capture and storage, as well as clean conversion technologies leading to substantial improvements in plant efficiency, reliability, and costs.

In the U.S.A., the world's first CCS power plant of 5 MW_e was put into operation in March 2005 by Clean Energy Systems, Inc. In addition, many new CCS projects have been presented and described at the 5th and 6th Annual Conferences on Carbon Capture and Sequestration in the U.S.A. (2006, 2007).

The aim of this book is to help all the people involved in the emerging clean fossil-fuel power industry. It is also intended to inform energy engineering students about the technology relevant to their future jobs to create and operate low- to zero emissions power generating plants, making the energy future clean, secure, and inexpensive.

In anticipation of the planned nuclear generating capacity reductions, the crucial question facing the European power generation industry today is: what type of power plants will be put into place in the future to fill the impending power vacuum?

1.4 UNAVOIDED CARBON CAPTURE AT ZEPP (ZERO EMISSION POWER PLANT)

Continued expansion of clean, renewable energy (wind, hydro, solar, wave, tidal motion, etc.) must be sustained. Photovoltaics are presently struggling to compete. Wind power is looking more promising and is growing fast with claims that the specific costs are declining. The main problem with renewables, however, is the very low primary energy current density presently associated with such technologies (i.e., too much land area is required to produce reasonable quantities of power). Land area is scarce in many places, especially in densely populated Europe. Renewables, even taken together, cannot meet the growing energy demands of industrial society. One promising approach, based on photosynthesis, is described in Chapter 8. The large figures for the installed capacity of wind farms should be read with caution. The availability rating of a wind farm is inferior to that of a fossil-fuel-fired power plant. When we compare the installed capacity of a fossil-fuel-fired power plant to that of a wind-powered plant, the quantity of annually produced electricity from wind is typically one fourth that produced from a fossil fuel plant of the same nominal power.

An example of the wrong application of very useful social energy is shown in Figure 1.5. It is a large placard by Greenpeace, one of many similar placards, at a demonstration on December 12, 2007, against a new coal-fired power plant near

FIGURE 1.5 Placard at a demonstration against a new coal-fired power plant in Neurath, Germany "Brown Coal Is Poison for the Climate." Note the opposite meaning of the word "Gift" in German and English. (Photo by author on Dec. 12, 2007 in Neurath, Germany.)

Neurath, Germany. There were some posters with even stronger wordings than "Coal is a Climate Killer." In Germany, where coal is the main energy resource, such pronouncements are very naïve. Actually, coal is not the killer, but rather the killer is the exhaust gases we call "emissions." The correct posters would read: "Emissions Kill the Climate." Before this, such declarations were not seen in the mass media or in public demonstrations.

Fossil-fuel-fired power plants produce energy at power densities typically about 100 times greater than renewable energy technologies. This simply means that fossil-fueled plants need much less land to produce the same amount of power. The mass media generally tend to lay the blame for anthropogenic pollution on the usage of fossil fuels. One often hears an appeal for a widespread ban on fossil fuel use. This appeal is totally misguided.

Fossil fuels themselves do not cause pollution. It is the way in which they are burned that is the problem. The desire by so many for a carbon-free energy supply is understandable, but it is not realistic in the short term. There may be carbon-free energy supplies in quantity in the distant future. The reality is that, over the next many decades, fossil fuels will dominate the energy mix (see Figures 1.1 and 1.2).

Fossil fuels are more affordable, more widely available, and more flexible than any other energy resource, and their economic advantages are not likely to change in this century.

The projected world energy development presented in Figure 1.1 and Figure 1.2 is not sustainable in view of current CO_2 emission stabilization objectives. The Kyoto protocol calls for the release of greenhouse gases within the EU needs to be reduced by 8% in 2012 with respect to 1990 levels. Even radical improvements in energy efficiency and the development and commercialization of numerous renewable energy technologies will not constitute adequate options to meet the stated CO_2 stabilization objectives. The ever-widening gap that is emerging with time between CO_2 stabilization requirements and the projected increase in fossil fuel usage can only be filled by new, carbon capture zero emissions, fossil-fuel-based technologies as clearly shown by the MIT forecast in Table 1.1. Bringing zero emissions, fossil-fuel-flexible, and highly efficient power generation technologies into commercial use is a critical task for the 21st century. In fact, the authors firmly believe that the urgent development of such technologies is the single most important issue related to fossil fuel use today.

In recent programs of leading energy supply and development institutions like the Department of Energy (DOE) in the U.S.A. ("Vision 21"), the zero emission power plant (ZEPP) concept has achieved top priority. Even so, there exists in the U.S.A., a sharp, irreconcilable controversy on the scale and cause of global warming and many doubt whether a large-scale investment in ZEPP technology is justified. There has been strong opposition to the implementation of such technology, which is not compatible with the MIT forecast.

The authors take the following position. Irrespective of global warming considerations, the ZEPP technology is needed to guarantee a basic human right — the right to breathe clean air. If we live on a river, nobody can make us drink dirty water, which is being polluted by a landlord upstream. The polluting landlord would be obliged to stop polluting the river. Why then, should we be forced to breathe dirty air, air polluted by power plants, boiler houses, cars, buses, and trains? The only realistic and viable solution is apparent: a concerted technological, environmental, and political will to facilitate and promote implementation of ZEPP technology.

ZEPP technology will:

- Provide affordable, clean power to meet expanding energy demand
- Solve critical environmental problems (eliminating carbon dioxide and pollutant emissions)
- Address energy security issues by supporting the use of diverse fossil fuels, including integrated coal gasification and pulverized coal combustion
- Ease the economic cost of sustainable energy supplies primarily by the use of cogenerated carbon dioxide for enhanced oil and gas recovery

Among the many schemes being developed for ZEPPs, one of the most promising is the combustion of gaseous fossil fuels in oxygen, moderated by recirculation of carbon dioxide, steam, or water. The combustion products drive turbines to produce power followed by heat recuperation and collection of carbon dioxide from the process in a liquid or supercritical state at high pressure. Recent achievements in oxygen separation with dense ceramic membrane technology are showing progress in this regard.

Every ZEPP produces electric power and, as a byproduct, a large quantity of carbon dioxide. The pressurized CO_2 may be sequestered deep underground or used for enhanced oil recovery, especially in largely depleted fields. The incremental oil production might be as high as 20% of the original oil in place. Recovery of coal bed methane from unmineable coal fields is of interest in a number of locations. Europe has large quantities of unmineable coal. Practical demonstrations of coal bed methane recovery by carbon dioxide injection have been made recently in Canada, the U.S.A., and Poland (by a joint European research group). To put the possibilities in perspective, an evaluation of world reserves of recoverable coal bed methane by U.S. geologist V. Kuuskraa, produced an estimate at 150 Gtoe (about 40 times the global annual oil production). German geologist D. Juch estimates the reserves of coal bed methane in the coal-rich Ruhr district as 0.4 Gtoe, about the total of oil reserves in Europe.

The fuel most easily utilized in ZEPPs is natural gas. It is relatively abundant but is also more costly than coal per unit of heat. The International Gas Union, at their June 2000 Congress, estimated that even a small part of known gas reserves would be sufficient for 200 years of global supply at current consumption rates. Some support for such optimistic views is provided by information related to the origin of hydrocarbon fuels.

1.5   THE ORIGIN OF HYDROCARBON FUELS

The known supplies of coal are vast but, in some scenarios, the amount of natural gas might be even greater. There is controversy among geologists about the quantity of available natural gas but observations of gas at depth, and in the form of hydrates, support the more optimistic gas reserves.

Natural gas consists mainly of methane. Methane can also be produced as biogas, typically from organic waste. Recent evidence suggests that much more methane exists than was previously suspected. Many authors have discussed the availability of methane hydrates, in which methane is "stored" inside a complex molecular structure. These hydrates could become an important source of methane in years to come. Deep gas pools also hold much unused methane. Expectations for exploration, development, and use depend upon the actual processes from which hydrocarbons originate.

The biogenic theory postulates that hydrocarbons originated from remains of former living matter. Such hydrocarbons would be located in sedimentary layers, not too deep under the Earth's surface. Such views have been popular up to now, especially in the West.

The abiogenic theory, first introduced by D. Mendeleev, postulates the production of hydrocarbon fuels by inorganic processes. In Mendeleev's 1877 paper, "On the Origin of Oil — The Petroleum Industry in Pennsylvania and in the Caucasus," (translated from Russian by N. Makarova) he stated:

> We can't suppose petroleum origin from ancient organism… we need to look for petroleum origin deeper in Earth crust than in the places where petroleum was discovered…. carbon metals exist inside of Earth … Iron and other metals with oxygen of water will produce oxides and hydrogen will be freed — partly as free hydrogen and partly with carbon … this is petroleum.

N. Koudriavtsev (1959, 1963) significantly developed this theory and E. Tchekaliuk (1967) offered a thermodynamic background for oil formation.

The best known recent advocate of abiogenic theory, Thomas Gold, discussed the production of hydrocarbon fuels simultaneously with other substances in the depths of Earth. Gold's view led him to predict rivers and seas of methane on the surface of cold planets where organic life never existed. Gold passed away in July 2004, 6 months before the European Space Agency's probe Huygens landed on Saturn's largest moon, the cold Titan (about −180°C), and produced photos of the rivers and seas of methane Gold had predicted. As we see in Figure 1.6, shot from an altitude of 8 km, and Figures 1.7 and 1.8, made after landing on January 14, 2005, the quantities of methane are astonishing. The distance from Titan to Earth is about 1.5 light-hours.

Starting from his essay "Rethinking the Origin of Oil and Gas" in *The Wall Street Journal* (June 8, 1977), Gold (1987, 1999) transformed his popular and technical papers into two books. His central ideas can be expressed in a few sentences. According to Gold, our planet is like a sponge, filled by primordial hydrocarbons. They were formed together with the other substances of the deep Earth about 4 billion years ago. Due to their much lower density than the surrounding porous rocks, an upwelling of hydrocarbons, mainly methane, takes place from a depth of hundreds of kilometers. This permanent seepage from below fills the known pools of oil and gas, and some eventually escapes into the atmosphere. Hydrocarbons lose hydrogen along the way from initial CH_4 finally to $CH_{0.8}$ (coal), and even in some

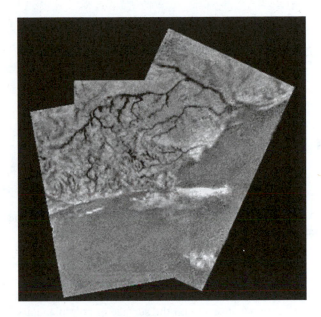

FIGURE 1.6 Rivers and sea of methane on the surface of Saturn's moon Titan seen from an altitude of 8 km. (*Source*: the European Space Agency (ESA)).

places and conditions to form pure carbon as diamond. In Gold's view, there is good reason to consider hydrocarbon fuels as an almost unlimited source of energy.

Gold gave strong arguments supporting his views, using his outstanding erudition in astrophysics, chemistry, and biology.

FIGURE 1.7 Endless sea of methane photographed from Titan's surface by probe Huygens, January 14, 2005. (*Source*: ESA.)

FIGURE 1.8 A lake of liquid methane surrounded by mountains of solid ice on Titan. (*Source:* Huygens probe, ESA.)

Some facts are impossible to explain by the biogenic theory:

- Methane rivers now observed on the surface of cold planetary bodies like Titan, where a possibility of organic life never existed.
- Tremendous methane outflow accompanying volcanic eruptions, together with hot lava.
- Refilling of depleted oil reservoirs, registered in many places, for instance near Eugene Island in the Gulf of Mexico (*The Wall Street Journal*, April 16, 1999).

The main argument of biogenic theory proponents is the existence in crude oil of some molecules that indisputably came from living matter because they manifest a chiral effect (strong rotation of polarized light). The presence of biogenic molecules is actually an important part of the abiogenic concept as well, but it turns the old logic upside down. It is not that bacteria produced the hydrocarbons, but that the primordial hydrocarbon "soup" provided food for bacteria. The total mass of the organic matter in these microbes is estimated to be hundreds of thousands of gigatons, much more than the organic mass of the surface biota.

Indeed, reasons exist to consider deep underground layers as the more probable place for the origin of life. Perhaps, ten planetary bodies in our solar system could produce suitable subsurface abodes for the same kind of life as we have within Earth. Any rocky planetary body at least as big as our moon might be expected to offer the requisite subsurface conditions of heat and upwelling hydrocarbons. An instrument placed on the moon by an Apollo mission detected gas particles of atomic mass 16, which could be methane. Panspermia (transfer of life) between planets of a planetary system would be a possibility, as a Martian meteorite suggests.

The triumph, not always recognized, of the abiogenic theory was the drilling experiment in the Siljan Ring in Sweden. Microbes were found at a depth of 6 km, living on oil and getting oxygen by reducing Fe_2O_3 to Fe_3O_4, thus leaving behind tiny particles of magnetite (a fraction of a micron in size) as the product of metabolism. Gold also observed that the liquid formed by these particles within the extracted oil sample had magnetic properties.

In addition to the drilling experiment in Sweden, a deeper one on the Kola Peninsula in Russia indicated, in complete accord with Gold's predictions, that at a depth of about 10 kilometers, the fractured rocks were filled with highly compressed methane. Former life could not have reached such a depth and produced such a large quantity of methane.

Some tectonic movements might be connected with upwelling of methane and accompanying gases such as hydrogen sulphide, which are in turn connected with earthquakes. An increase of surface gas emissions before earthquakes is often sensitively detected by rats and other animals, which manifest unusual behavior prior to earthquakes. Such behavior is documented by numerous earthquake eyewitnesses.

Abiogenic theory also offers an explanation of the formation of metal deposits in the deep Earth. Reaction of upwelling hydrocarbons also produces organo-metallic compounds along the way.

Gold (1987) respectfully acknowledges his predecessors in abiogenic theory, in particular the Russian and Ukrainian scientists, Mendeleev, Koudriavtsev, Tchekaliuk, and Kropotkin. Kropotkin provided Gold much geological information.

Liquid methane cannot exist on Earth's surface due to its temperature but, at the high pressure in the upper mantle of the Earth, methane could exist in a supercritical state. One of the last arguments of biogenic advocates is that it is possible to find abiogenic methane or other hydrocarbons in the Earth's crust, but the quantity is rather small, and consequently, not commercially useful. From photographic images, we can see a very large quantity of methane exists on Titan, but it is not realistically useful for Earth because Titan is 1.5 light-hours away. The large quantities of methane seen in volcanic eruptions, as methane hydrates, or deep drill cores in the Kola Peninsula, are still not commercially available either. Exploration for large quantities does not violate mass conservation or other physical laws. Substantial quantities of methane exist on Earth. In the authors' view, the challenge is for technology to locate, recover, and use it.

Evidence that methane can be produced by inorganic processes has been shown experimentally by Scott et al. (2004), who formed methane from FeO, calcite ($CaCO_3$), and water at pressures between 5 and 11 GPa and temperatures ranging from 500 to 1500°C. Other relevant data come from work under hydrothermal conditions (Horita and Berndt, 1999). These experiments, backed by calculation, demonstrate that hydrocarbons may be formed in the upper mantle of Earth and confirm the general view formulated in a book by Korotaev et al. (1996), where P. N. Kropotkin compiled substantial data on abiogenic methane.

If sources of methane abound, the main restrictions on its use are pollution of the atmosphere by exhaust gases and the amount of oxygen available for combustion in the atmosphere. High-temperature, high-pressure ZEPPs offer the possibility of using currently available methane stores efficiently without pollution. Oxygen depletion of the atmosphere is unlikely to be a problem in the current century. The total mass of oxygen in the atmosphere is about 10^{15} tons, while the current annual consumption of oxygen is only about three 10^{11} tons. Any problems caused by this 0.03% annual usage of oxygen have yet to become apparent. We may expect that when, or if, a decrease in oxygen levels ever becomes apparent, solar, nuclear, or other alternatives will dominate as sources of primary energy.

1.6 THERMODYNAMICS OF A REACTION WITH METHANE FORMATION OF CO_2 AND FAYALITE

An example of methane origination in the Earth's crust in a man made technology is calculated below. It might be a hint to previously mentioned natural methane origination according to abiogenic theory.

Consider the simplified chemistry of carbon reactions, Equation (1.1) in a ZEPP (above the line) and in an artificial geochemical reactor deep underground (below the line)

$$
\begin{array}{c}
CH_4 \;+\; 2O_2 \;\Rightarrow\; CO_2 \;+\; 2H_2O \;+\; Power \\
\uparrow \qquad\qquad\qquad\qquad\qquad \downarrow \\
CH_4 \;+\; 4Fe_3O_4 \;+\; 6SiO_2 \;\Leftarrow\; CO_2 \;+\; 2H_2O \;+\; 6Fe_2SiO_4
\end{array}
\tag{1.1}
$$

The four arrows show a closed cycle of carbon. *Carbon plays the role of an energy carrier from fayalite to methane and then to the power plant.* At realistic conditions of temperature and pressure, some hundreds of degrees Celsius and some hundreds of bar, the reaction is exothermic and may be self-sustaining.

Below are reproduced calculations by Akinfiev et al. (2005) on formation of methane by a reaction of carbon dioxide with fayalite, an abundant mineral (see Wooley, 1987). *Fayalite becomes a fuel, with methane being the energy carrier. Viewed in this way, methane is a renewable energy source*, as theorized by Yantovsky (1999). Table 1.2 shows the properties of the reactants.
For the reaction

$$
6Fe_2SiO_4 + 2H_2O + CO_2(g) = 4Fe_3O_4 + 6SiO_2 + CH_4(g) \tag{1.2}
$$

properties of both liquid (l) and gaseous $H_2O(g)$ were calculated using Hill's (1990) equation of state. At the temperature 250°C and pressure 200 bars, we have

$$
\Delta_r G_{250°\ C/200\ bar} = 4 \cdot (-1,053,163) + 6 \cdot (-868,582) + (-95,370) - 6 \cdot (-1,424,175)
$$

$$
-2 \cdot (-257,755) - (-445,257) = -13,697\ J \tag{1.3}
$$

$$
\Delta_r H_{250°\ C/200\ bar} = 4 \cdot (-1,078,057) + 6 \cdot (-898,100) + (-65,345) - 6 \cdot (-1,446,764)
$$

$$
-2 \cdot (-268,156) - (-383,990) = -165,287\ J \tag{1.4}
$$

The last value means that the reaction is exothermic.

TABLE 1.2
Properties of Selected Reactants

Substance	Enthalpy h, J/mole	Gibbs Free Energy g, J/mole
Fayalite	−1,446,764	−1,424,175
H₂O	−268,156	−257,755
CO₂(g)	−383,990	−44,5257
Magnetite	−1,078,057	−1,053,163
Quartz	−898,100	−868,582
CH₄(g)	−65,345	−95,370

Source: Johnson, J.W. et al., 1992.

The decimal logarithm of the thermodynamic constant of reaction for Equation (1.2) is

$$\lg K_a = \lg\left(\frac{f_{CH_4}}{f_{CO_2}}\right) = -\frac{\Delta_r G}{2.303 \cdot RT} = -\frac{-13\,697}{2.303 \cdot 8.314 \cdot 523.15} = 1.37 \quad (1.5)$$

Use of the Redlich-Kwong equation of state shows that the fugacity coefficients of the gases are close to one ($\gamma_{CO_2} = 0.87$, $\gamma_{CH_4} = 1.01$ at the T and P specified). This means

$$\log K_a = 1.37 = \log(P_{CH_4}/P_{CO_2}), \quad (1.6)$$

$$\frac{P_{CH_4}}{P_{CO_2}} = 23.3 \quad (1.7)$$

The last value brings the conversion factor of CO_2 into CH_4 close to 96%.

We are unaware of experiments on the interaction of fayalite and carbon dioxide. However, the Albany Research Center (O'Connor et al., 2000) extensively examined a similar reaction in an autoclave with comminuted (37–106 micrometers) forsterite Mg_2SiO_4. Chemical energy of the forsterite is 256.8 kJ/mole. The reaction products are magnesite $MgCO_3$ and silica or silicic acid. The pressure was 117 to 127 Atm and temperature 185 to 188°C. Test times spanned 3 to 48 hours. A large increase of the reaction rate was observed when carbon dioxide was in a supercritical state. Similar experiments with comminuted fayalite might shed light on the kinetics and applicability of reaction (1.2) for production of methane.

Several recent diverse observations and experiments favor abiogenic theory and the possibility of finding deep in Earth what we see on the surface of Titan. Moreover, calculations show that the reaction between fayalite and CO_2 is exothermic and might be self-sustaining. A valuable next step would be experimental demonstration of the reaction of comminuted fayalite with carbon dioxide and water at conditions such as those mentioned above. Better fuels than fayalite might even be discovered.

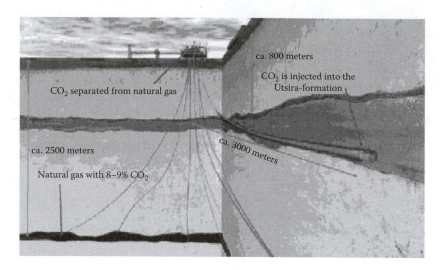

FIGURE 1.9 The first in the world sequestration of carbon dioxide in depth underground. (*Source: Statoil Magazine*, Special Issue on CO_2, Aug. 15, 2007, p. 11. //www.snohvit.com/ statoilcom/SVG00990.NSF/Attachments/co$_2$MagasinAugust2007/$FILE/CO$_2$_eng.pdf. Accessed Nov. 20, 2007.)

The search for methane in Earth and resolution of its origins deserve more research efforts than ever before. Quite recently, a big discharge of deep methane has been registered in the Barentz Sea and East Siberian Sea. From the point of global warming, it is a dangerous event, as methane is a greenhouse gas. But it manifests the existence of deep gas in accordance with Gold's theory. In any case, we should pay tribute to Thomas Gold for his forecasts.

1.7 EMERGING TASK — THE SEQUESTRATION

Along with searching for fuel in depth is emerging the other task for geologists: finding the appropriate place to sequester the products of the fuel combustion. This problem has been discussed at many international meetings over the last two decades. Now almost unanimous agreement of geologists is that sequestration of carbon dioxide (with dissolved contaminants) in the depth of more than 800 m is quite possible. Figure 1.9 illustrates the biggest such enterprise, created in Norway. Up until now, there appeared to be no hint on the escape of injected gas.

REFERENCES

Annon., World Energy Outlook. 2004. www.iea.org/textbase/npdf/free/2004.
Annon., WEU. 2006. Sample Graphics. www.iea.org.
Akinfiev, N. et al. 2005. Zero Emissions Power Generation with CO_2 Reduction by Fayalite, *Int. J. Thermod.* 8 (3):155–157.

Collins, W. et al. The Physical Science Behind the Climate Change. *Scientific America*, August/2007, p. 67.

Gold, T. 1987. Power from the Earth: Deep Earth Gas — *Energy for the Future*. London: Dent & Sons Ltd.

Gold, T. 1999. *The Deep Hot Biosphere*. New York: Copernicus, Springer V.

Hill, P.G. 1990. A unified fundamental equation for the thermodynamic properties of H_2O. *J. Phys. Chem.* (19): 1233–1247.

Horita, J. and M.E. Berndt. 1999. Abiogenic methane formation and isotopic fractionation under hydrothermal conditions, *Science*, 285 (5430): 1055–1057.

Johnson, J.W., Oelkers, E.H., and H.C. Helgeson. 1992. SUPCRT92: A software package for calculating the standard molal thermodynamic properties of minerals, gases, aqueous species, and reactions from 1 to 5000 bar and 0 to 1000°C. *Comp. Geosci.* 18: 899–947.

Katzer, J. et al. 2007. *The Future of Coal*, MIT, http://web.mit.edu/coal/The_Future_of_Coal_Summary_Report.pdf (accessed on Oct. 11, 2007).

Korotaev, Y. et al. 1996. Methane's epoch is not a myth, but a reality, in *Proc. of Int. Fuel-Energy Assoc.* Moscow (in Russian).

Koudriavtsev, N. 1959. *Oil, Gas and Solid Bitumen in Metamorphic Rocks*. Leningrad: Izd. *Gostoptechizdat* (in Russian).

Koudriavtsev, N. 1963. *Genesis of Oil and Gas*. Moscow: Izd. *Nedra* (in Russian).

Marchetti, C. 1979. Constructive solutions to the CO_2 problem, in *Man's Impact on Climate*, Elsevier: New York.

O'Connor, W.K., Dahlin, D.C., Turner, P.C., and R.P. Walters. 2000. Carbon Dioxide Sequestration by Ex-Situ Mineral Carbonation. *Technology* (7S): 115–123.

Scott, H., et al. 2004. Generation of methane in the Earth's mantle: In situ high pressure-temperature measurements of carbonate reduction. *Proc. Nat. Acad. Sci. U.S.A.*, 101 (39): 14023–14026.

Tchekaliuk, E. 1967. *Oil of Upper Mantle of the Earth*. Kiev: Izd. *Naukova Dumka* (in Russian).

Wooley, A. 1987. Alkaline Rocks and Carbonates of the World, Part 1, in British Museum (Natural History), London.

Yantovsky, E. 1999. On the geochemical hydrocarbon reactor concept, in *Proc. Fifth Int. Conf. on Carbon Dioxide Utilization* (ICCDU-5). Ges. Deutscher Chemiker, Frankfurt a/M, Sept. 5–10, Karlsruhe, Germany, p. 51, 224–225.

2 Cycles Review

2.1 CARBON CAPTURE METHODS

We realize there are many different readers of this book. Some are not interested in details and only wish to know which cycles exist and their main features. For these readers the following is a short history and the state-of-the-art. Later, the main features in our review are discussed in detail.

Fossil fuels are well understood by the power industry, and are still relatively cheap and abundant. There is at least as much oil still available in the world as has been burned since the industrial revolution, and far more methane available than has so far been consumed. For good or bad, fossil fuels will remain an integral part of the energy production mix for many decades to come. Unfortunately, combustion of these fuels produces carbon dioxide, the main contributor to global warming.

Zero emission power plants (ZEPPs) offer a method of producing energy from fossil fuels without emitting carbon dioxide. These plants could replace decommissioned power plants, including nuclear plants.

There are many technologies utilizing zero emissions combustion of fossil fuels, not all of which involve power production. The interested reader is referred to the Zero Emission Technologies for Fossil Fuels: Technology Status Report, published by the International Energy Agency (IEA, 2002).

ZEPPs must follow the law of conservation of mass. Every atom of fuel or oxidizer entering a plant must leave as ash, emission, or effluent. In the ZEPP, all the combustion products are converted to liquid form. In short, the problem of zero emissions translates into the conversion of gaseous emissions into liquid effluents.

Many methods of carbon capture attempt to clean the exhaust gases after combustion by using absorption or adsorption, or to extract carbon from the fuel. All of these attempts lead to large mass exchangers. Some produce impure streams of carbon dioxide and none remove all of the carbon dioxide from the exhaust. In short, these are not zero emissions power plants, and as such are beyond the scope of this book. The only true zero emissions form of combustion in existence today is precombustion gas separation. This procedure is combustion of fuel using oxygen instead of air, usually diluted with either CO_2 or water vapor. This produces an exhaust stream containing only carbon dioxide and water vapor, which are easily separated by condensing out the water vapor. The pure carbon dioxide stream is then compressed and condensed to produce a manageable effluent of liquid CO_2, which can be sold or sequestered. This approach is often called *oxyfiring* or *oxyfuel*. Figure 2.1 shows a schematic of the various methods of carbon capture. Oxyfiring is the left downward branch in the picture.

The demand for CO_2 in world industry was evaluated as 1.6% of the CO_2 released to the atmosphere by power plants (Pechtl, 1991). This implies that if zero emissions

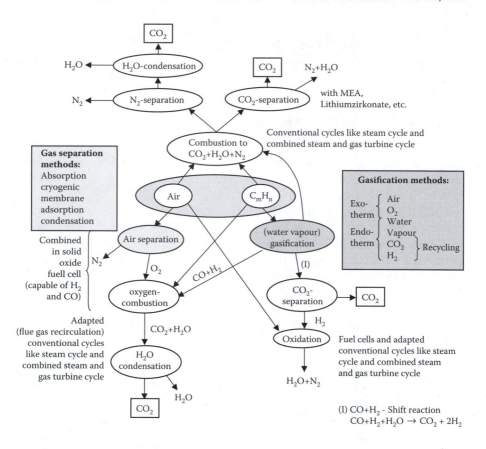

FIGURE 2.1 The various methods of carbon capture (Leithner, 2005).

power generation was to be implemented on a global scale, the sequestration of CO_2 would be unavoidable.

The phrase *zero emissions* can be challenged for two reasons. First, some cycles convert the liquid water back to water vapor and emit it into the atmosphere. In this case, the cycle is not strictly zero emissions; however, the water vapor is not a pollutant or greenhouse gas. Second, some cycles intend for the carbon dioxide to be of high purity, requiring removal of contaminants. Some carbon dioxide will probably escape with these contaminants. If the carbon dioxide is destined for storage, however, it does not require a high purity, and the contaminants can be stored along with the carbon dioxide. The authors acknowledge that there may be some cases in this chapter in which the phrase *zero emissions* is not strictly accurate. All values given in this chapter are taken from the referenced papers and are subject to the varied assumptions and, in some cases, mistakes in those papers. They were calculated using different assumptions, models, and boundary conditions. This chapter is only a history of zero emissions cycles, not a comparative evaluation of the different cycles.

Therefore, the values given here, particularly the efficiencies, should be considered accordingly.

The compression of the carbon dioxide reduces the efficiency of the plant, but the reduction may be acceptable given the damaging effects of carbon dioxide emissions. Production of oxygen further reduces the efficiency. The most mature method of oxygen production is cryogenics, an energy intensive process involving freezing air. Ion transport membranes offer a much more efficient method of producing oxygen, with the result that many ZEPP cycles incorporate these membranes. This chapter also presents the history and current state of the art of these cycles.

2.2   EARLY ATTEMPTS

To our knowledge, the first mention of a zero emissions power unit was documented by Degtiarev and Gribovsky (1967) (Figure 2.2). This system integrates air separation, power generation, combustion of a gas in a mixture of CO_2 and oxygen, and

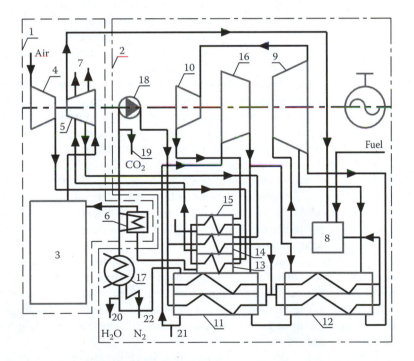

FIGURE 2.2 Reproduction of the schematics of Degtiarev and Gribovsky (1967). Outline of carbon dioxide zero emissions power plant by V.L. Degtiarev and V.P. Gribovsky of July 28, 1967: 1-oxygen plant (ASU), 2-carbon dioxide power plant, 3-rectifier, 4-air compressor, 5-oxygen compressor, 6-air cooler, 7-compressed oxygen extraction, 8-high-pressure combustion chamber, 9-main turbine, 10-auxilliary turbine, 11-combined regenerator, 12-main regenerator, 13, 14, 15-regenerative heater, 16-compressor, 17-condenser, 18-pump, 19-liquid CO_2 release, 20, 21-water release, 22-nitrogen release.

production of liquid CO_2. The only emission is the cold nitrogen from air separation. The aim of the unit is cogeneration of electric power and carbon dioxide for industry. Both authors were students of Professor D. Hochstein at the Odessa Polytechnic Institute, who proposed high pressure carbon dioxide as the working fluid in a Rankine cycle (Hochstein, 1940). At that time, the greenhouse effect was ignored.

Recently, CO_2 as a working substance was proposed by Tokyo Electric (Ausubel, 2004). This is a gas turbine cycle with the exhaust gases condensed, compressed, heated, and returned to the combustion chamber. The cycle called for 1500°C and 400 Atm at the turbine inlet, which seems rather optimistic. This cycle owes much to Professor Hochstein's work.

Marchetti (1979) proposed combustion of fuel in a CO_2/O_2 mixture, followed by CO_2 sequestration in the ocean. The mass balance for fuel, oxygen, and CO_2 is given.

The concept of total emission control combined with enhanced oil recovery was described by Steinberg (1981). His history of carbon mitigation technologies (Steinberg, 1992) describes his pioneering work in this area from 1981 to 1990.

Another history of various zero emissions cycles was presented by Yantovsky and Degtiarev (1993). For each of the two possible recirculating substances (H_2O and CO_2), the external combustion (Rankine cycle) or internal combustion (oxyfired) branches are identified. The latter is divided into many particular cases. Known data on cycle efficiencies is compared on a graph, which reveals a much higher efficiency using CO_2 recirculation as opposed to H_2O recirculation.

Rankine cycle using zero emissions combustion of coal powder was well documented in Argonne Lab by Wolsky (1985), and Berry and Wolsky (1986). Coal fired zero emissions combustion was also investigated by Nakayama et al. (1992), who concluded "O_2/CO_2 combustion process will provide an effective pulverized coal-fired generating system for CO_2 recovery." Only cryogenic oxygen was considered at that time, which resulted in a cycle that was not economically viable. However, more recently, investigation into a zero emissions coal cycle using ion transport membranes has begun at the Technical University of Aachen (Renz et al., 2005). This cycle, named Oxycoal-AC, is presented as having 41% efficiency for a 400 MW plant. This cycle is essentially the same as the Milano cycle (Romano et al., 2005), which also claims a 41% efficiency.

Pak et al. (1989) gives a description and schematic of a ZEPP cycle with CO_2 recirculation. However, the need to deflect combustion-born water from the cycle is missed.

Lorentzen and Pettersen (1990) presented a ZEPP cycle in which gas is combusted in a CO_2/O_2 mixture. They not only present the scheme, but also the T-s diagram, on which the thermodynamic losses are clearly indicated. These authors have much experience in the use of CO_2 in refrigerators.

Pechtl (1991) considered a ZEPP cycle with CO_2 recirculation and gave some figures, subsequently confirmed by other authors. If the efficiencies of an ordinary 500 MW coal-fired power plant and an equivalent ZEPP are compared, the efficiency drops from 38.9 to 36%. The liquefaction of CO_2 takes 5.3% of the generated power.

Yantovsky (1991) presented in some detail a schematic for a ZEPP with combustion of natural or coal-derived gas in an O_2/steam mixture, with triple turbine

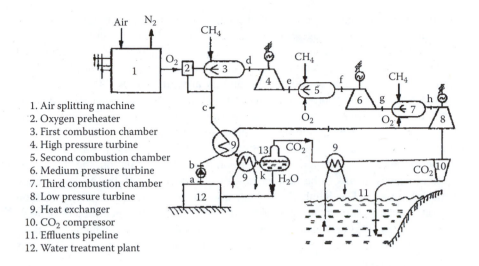

1. Air splitting machine
2. Oxygen preheater
3. First combustion chamber
4. High pressure turbine
5. Second combustion chamber
6. Medium pressure turbine
7. Third combustion chamber
8. Low pressure turbine
9. Heat exchanger
10. CO_2 compressor
11. Effluents pipeline
12. Water treatment plant

FIGURE 2.3 Power plant without exhaust gases (Yantovsky, 1991).

expansion, CO_2 separation for sequestration, and water recirculation. Estimation of the cycle efficiency by means of a *T-s* diagram, at a temperature of 750°C before each of the three turbines, gives an efficiency of 37%. In this paper, the practice of emitting carbon dioxide into the atmosphere was likened to the medieval practice of emptying chamber pots into the streets and neighbors' gardens.

The cycle presented in that paper and shown in Figure 2.3, comprises an air splitter, from which nitrogen is returned to the atmosphere and oxygen is diluted with recirculated water vapor and burned with methane in a combustion chamber, before entering the turbine. The exhaust is cooled, the water separated out for recirculation, and the CO_2 compressed for sequestration. The turbine inlet temperature of 750°C was based on a turbine without blade cooling. Yantovsky et al. (1992) presented a computer simulation of this ZEPP at higher turbine temperatures. At the highest turbine inlet temperature of 1300°C before the turbines, the efficiency does not exceed 40%. The low efficiency is caused by the very high latent heat of water and the inability to recuperate the large enthalpy of the condensing steam. As water is the recirculated substance, this turbine inlet temperature may be optimistic.

2.3 INDUSTRY FIRST BECOMES INTERESTED

Five years after Yantovsky's paper, Clean Energy Systems, Inc. (CES) patented a similar cycle, based on rocket engine technology in which fuel, oxygen, and water enter a combustor, (Beichel, 1996). This produces a very high temperature jet of 90% steam with 10% CO_2. Further stages dilute the mixture with more water, increasing the mass flow rate and reducing the temperature. As turbine technology improves, these stages may be removed, increasing the turbine inlet temperature and allowing the plant to increase in efficiency as technology improves. The turbine exhaust is cooled in stages, condensing out the water for recirculation, and the CO_2 is compressed for sequestration. The oxygen is produced by an unspecified air separator, which may be cryogenic or ITM based. A five-megawatt demonstration plant in Kimberlina, California, began operating in March 2005 (Figure 2.5). The oxygen is currently produced externally, but an onsite air separation plant is planned. The 10-megawatt combustor has been successfully tested. The Kimberlina plant is the first zero emissions power plant in the world. The cycle is shown in Figure 2.4. (See detailed recent descriptions in www.cleanenergysystems.com.)

Due to the use of water rather than carbon dioxide recirculation, the thermodynamic efficiency seems to be limited to 40%. In recent years, the cycle has been further developed. The Kimberlina plant, always intended to be a demonstration plant rather than a commercial enterprise, will soon be joined by a commercial 50 MW plant in Norway. This new plant will use the developed cycle, in which nitrogen from the air separator is used to provide power. This new scheme (Marin et al., 2005) is shown in Figure 2.6.

The use of nitrogen offers a significant benefit: the high pressure nitrogen is heated by the combustion gases and produces work. This combats the inability to recuperate the latent heat of water, and is a real achievement in the development of the CES cycle.

Some technical data for this final cycle, called the *Zero Emissions Norwegian Gas* (ZENG) cycle, are shown in Table 2.1. The efficiency of 45% is not too high; some cycles using CO_2 as a working substance have higher predicated efficiencies, but these cycles usually assume very high turbine isentropic coefficients.

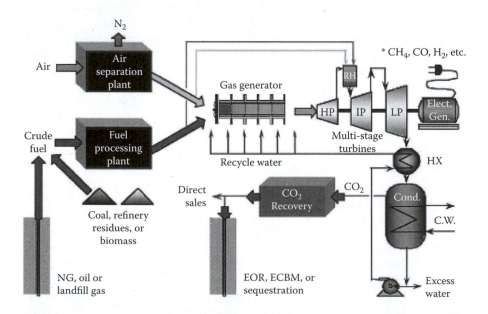

FIGURE 2.4 Clean energy systems cycle (Anderson et al., 2002).

FIGURE 2.5 First in the world ZEPP of 5 MW in Kimberlina, California, put in operation in March 2005 by CES with the cycle in Figure 2.4 (http://www.cleanenergysystems.com/technology.html).

The last achievements by CES, working with such industrial giants as Siemens and General Electric, reveal great advances in oxi-fuel CES cycles (Anderson et al., 2008). Now the intermediate turbine is replace by existing gas turbines. For modest T1T 760°C with the turbine J79 of GE the net power with all three turbines was 60 MW at 30% efficiency. By 927°C power is 70 MW at 34%. They are planning to get by 1260°C the power up to 200 MW at 40–45%.

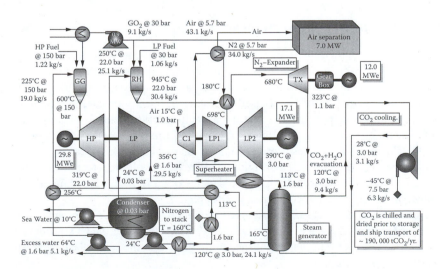

FIGURE 2.6 The ZENG project cycle (Marin et al., 2005).

TABLE 2.1
Technical Data for the ZENG Cycle

Optimized Cycle Summary Data

Thermal power input	114 MW
Gross power output	58.6 MWe
Parasitic power	8.1 MWe
Net power	50.5 MWe
Overall cycle efficiency	~45%
Fuel consumption	8,210 kg/h
Oxygen consumption	32,760 kg/h
Cooling water flow (total)	~4,300 m³/h
Excess water production	19.0 m³/h
HP turbine inlet pressure	150 bar
HP turbine inlet temperature	600°C
HP turbine exhaust temperature	~320°C
IP turbine inlet pressure	22.0 bar
IP turbine inlet temperature	698°C
IP turbine exhaust temperature	~390°C
CO_2/steam condenser pressure	3.0 bar

LP Steam Rankine Cycle

LP turbine inlet pressure	1.6 bar
LP turbine inlet temperature	356°C
LP turbine exhaust temperature	24°C
Steam condenser pressure	0.03 bar

Source: Marin, O. et al., 2005.

2.4 CONTINUED DEVELOPMENT

Bolland and Saether (1992) presented a paper documenting a ZEPP combined cycle—a gas turbine with combustion in an O_2/CO_2 mixture and a bottoming ordinary Rankine cycle. The production of oxygen and compression of CO_2 combine to reduce the efficiency from 56% (based on a state-of-the-art air-based cycle) to 41%. The same work also presents a single stage steam turbine cycle with combustion in an O_2/steam mixture. The maximum efficiency of this cycle at 1550 K is given as 38.5%. This paper also gives useful economic data on equipment costs.

De Ruyck (1992) proposed an original ZEPP cycle involving water evaporation in a mixture with CO_2. Extremely high efficiencies of up to 57% were claimed. These figures, however, were not confirmed in later papers.

Holt and Lindeberg (1992) considered an integrated complex comprising a ZEPP with enhanced oil recovery. They concluded that two-thirds of the CO_2 produced by combustion in a ZEPP might be returned underground to the same place from where the fuel was extracted.

Van Steenderen (1992) considered the combined gas/steam ZEPP in more detail. At 20 bars and 1050°C at the inlet to the turbine, an efficiency of 44% is reported.

Yantovsky et al. (1993, 1994a) and Wall et al. (1995) present a 10 MW ZEPP cycle with liquid CO_2 cogeneration, the latter being used to enhance oil recovery. The efficiency is presented as 48% at a turbine inlet temperature of 1000°C and pressure of 40 bars. The Aker Company in Norway began a similar project 5 years after these papers.

A highly efficient ZEPP cycle with CO_2 recirculation and gas combustion in an O_2/CO_2 mixture is described in detail by Yantovsky et al. (1994b). This is the CO_2 *Prevented Emission Recuperative Advanced Turbine Energy* (COOPERATE) cycle. The efficiency range is given as 46.9 to 55.2% for turbine inlet temperatures between 950 and 1350°C and pressures between 4 and 240 bars.

Discussions with turbine manufacturers established that the temperature and pressure before the first turbine in the COOPERATE cycle were not feasible in the foreseeable future. This led to further development of the cycle by Yantovsky (1994c), resulting in an efficiency of 50% based on more realistic turbine inlet states of 600°C at 240 bars and 1300°C at 40 bars (Figure 2.7 and Figure 2.8). This highly efficient, realistic cycle (COOPERATE demo) is described as quasi-combined as it consists of two parts: a high pressure Rankine cycle using CO_2 and a low pressure Brayton cycle using the same CO_2. A comprehensive description of almost all such zero emissions cycles can be found in the book by Göttlicher (1999). The COOPERATE cycle belongs to "Process Family II" in the book.

Yantovsky (1996) compares the COOPERATE cycle to a standard combined cycle, as shown in Table 2.2. The payback period was estimated as 3 years for the COOPERATE cycle if a fuel with a negative price, such as used lubricant, is used.

In the U.S.A., about one Mton/year of used lubricant is available. In the paper, the benefits of enhanced oil recovery are described, along with storage of carbon dioxide in brine.

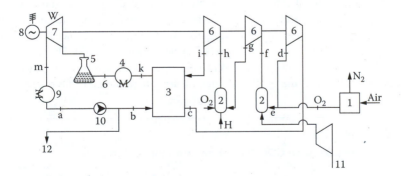

FIGURE 2.7 The COOPERATE-demo cycle (Yantovsky, 1994c). 1-Air Separation unit; 2-Combustion chamber; 3-Recuperator; 4-Cooling tower; 5-Water separator; 6-Turbine; 7-Intercooled multi-staged compressor; 8-Generator; 9-CO_2 condenser; 10-CO_2 pump; 11-Fuel; 12-Depleted well or other CO_2 storage.

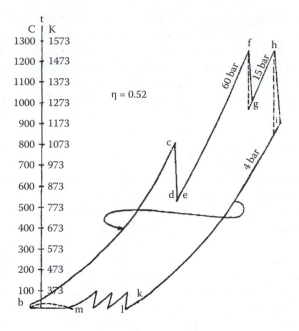

FIGURE 2.8 *T-s* diagram of COOPERATE demo cycle (Yantovsky, 1994c).

Using data on the amount of brine in the hydrolithosphere, a storage capacity of two million Gton of CO_2 in the brine solution is estimated. This could protect the atmosphere for the foreseeable future if to stop atmospheric emissions, redirected to hydrolithosphere. It should be noted that this figure is an upper limit and local restrictions should be taken into account in individual case studies.

A problem with the COOPERATE cycle is the noncondensable gases in the CO_2 condenser. As a radical remedy, it was proposed that CO_2 condensation could be avoided by compressing the CO_2 flow immediately after exiting the cooling tower, without allowing the compression process to cross the saturation line. This version of the COOPERATE cycle is the MATIANT cycle (Mathieu, 1998). Detailed calculations of the various versions of the MATIANT cycle (Mathieu et al., 1999) show that the loss of efficiency resulting from cryogenic air separation and CO_2 compression is

TABLE 2.2

Comparison of COOPERATE and Combined Cycle

	Efficiency	Cost of Electricity [€c/kWh]	CO_2 emissions [g/kWh]
Standard cycle	52.2%	4.00	360
COOPERATE	54.3%	5.55	0

Source: Yantovsky, E., 1996.

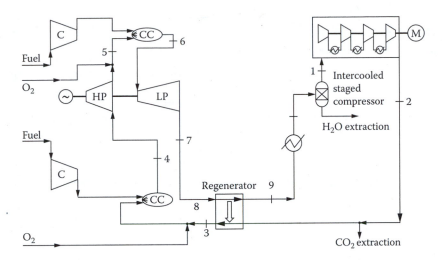

FIGURE 2.9 The MATIANT cycle (Mathieu, 1998).

about 11.5 to 14.5 percentage points, compared to a state-of-the-art cycle operating between the same thermodynamic parameters. The cycle involves staged combustion with a 2-stage expansion, and is shown in Figure 2.9 and Figure 2.10, along with some technical information, given in Table 2.3.

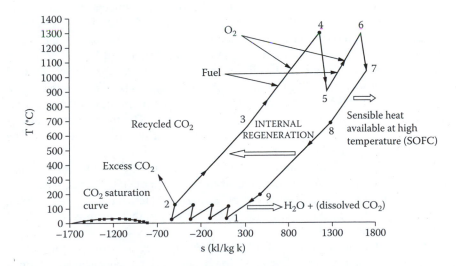

FIGURE 2.10 T-s diagram of the MATIANT cycle (Mathieu, 1998). 1–2: Intercooled staged compressor; 2–3: Upper-pressure part of the regenerator; 3–4: High-pressure combustion chamber; 4–5: High-pressure expander; 5–6: Low-pressure combustion chamber; 6–7: High-temperature heat exchanger; 8–9: Regenerator; 9–1: Water cooler/separator.

TABLE 2.3

Operating Parameters of the MATIANT Cycle

Upper-cycle pressure	60 bar	Pinch-point at the regenerator outlet	20°C
Lower-cycle pressure	1 bar	Maximum inlet temperature in the regenerator	700°C
Pressure drop in the combustion chamber	3%p_{su}	Expanders inlet temperature (TIT)	1,300°C
Isentropic effectiveness of the three expanders	0.87	Lower cycle temperature	30°C
Isentropic effectiveness of the O_2 compressors	0.75	Isentropic effectiveness of the fuel compressor	0.75

Isentropic effectiveness of the intercooled CO_2 compressor: 0.85 for the first three stages, 0.8 for the last one.

Source: Mathieu, P., 1998.

Based on the data in Table 2.3, the cycle efficiency is around 45% when the fuel is natural gas, the turbine inlet temperature is 1300°C, and the exhaust gas temperature is limited to 700°C. As the exhaust gases are cooled by a steam reheat cycle, the efficiency climbs to 49%.

To efficiently use the heat at a higher temperature than 700°C, a solid oxide fuel cell (SOFC) was integrated into the cycle between points 7 and 8, in Figure 2.9 (Mathieu and Desmaret, 2001).

The isentropic effectiveness in the Matiant cycle are quite conservative and the results of the calculations are reliable and comply with technical limitations on the temperature at turbine inlet and exhaust. When the upper pressure changes from 140 bars (at 1200°C) to 220 bars (at 1400°C), the cycle efficiency increases from 44.3 to 46.05%. A coal-fired cycle with integrated gasification (IGCC-MATIANT) has also been developed (Mathieu and van Loo, 2005). At 1250°C and 120 bars, the calculated efficiency is 44.8%, which is rather high for a coal-fired ZEPP.

Ruether et al. (2000) described an integrated system with "oxygen-blown dry coal entrained gasification providing fuel to the MATIANT cycle. Oxygen for both the gasifier and the MATIANT cycle is prepared by use of an ion transport membrane (ITM) instead of a conventional cryogenic air separation unit." The thermal efficiency of the overall cycle is 43.6% of the higher heating value of the coal, and 99.5% of the carbon dioxide produced is captured.

The Graz cycle was first introduced by Jericha et al. (1995) and has been continually developed by researchers at the Institute of Thermal Turbomachinery and Machine Dynamics at Graz University of Technology (Jericha et al., 1995, 2000, 2004). It is similar to the Clean Energy Systems cycle in that steam is recirculated to the combustion chamber; however, it is more complex than the CES cycle. It was developed as an adaptation of a hydrogen/oxygen cycle published by Jericha (1985). The cycle was also developed for a coal-derived syngas plant (Jericha et al., 2000). The exhaust gas (80% steam and 20% carbon dioxide) powers a high-temperature turbine, after which about half is cooled, compressed and re-enters the combustion chamber, while the rest enters an intermediate pressure turbine. After the intermediate stage, a portion is bled off and the water condensed out. The rest enters

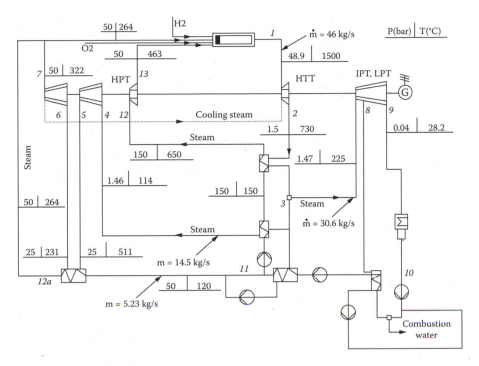

FIGURE 2.11 The original Graz cycle (Jericha et al., 1995).

a low-pressure turbine and is then cooled, with the water condensed out. The CO_2 is captured at the pressure of the intermediate turbine (atmospheric pressure). The water captured at low pressure is pumped to a very high pressure and heated. It then enters a steam turbine before returning to the combustion chamber. An efficiency of 56.8% of higher heating value was claimed for this cycle; however, the cycle assumes a supply of pure oxygen, and the carbon dioxide is provided at atmospheric pressure. The cycle is shown in Figure 2.11.

The Graz cycle has been developed in a practical manner, making use of the expertise available within the group in gas and steam turbine and heat exchanger design. This development led to a majority CO_2 flow cycle (Jericha et al., 2004; Heitmeir et al., 2003). This led to a significant body of work on the development of a 75% CO_2, 25% steam turbine. However, further development returned to a majority (77%) steam cycle, for which an efficiency of 70% was claimed, falling to 57% when oxygen production and liquefaction of carbon dioxide was taken into account (Sanz et al., 2004). Based on these results, Statoil became interested in the project and initiated an investigation into the Graz cycle. This resulted in a realistic efficiency of 52.6%, for a natural gas fired cycle, which takes into account not only oxygen supply and compression of carbon dioxide to 100 bars, but also mechanical, electrical, and auxiliary losses (Sanz et al., 2005). Ignoring these last three losses, the efficiency would be 54.6%. This latest incarnation of the cycle is shown in Figure 2.12. The flow is 75% steam, 25% carbon dioxide.

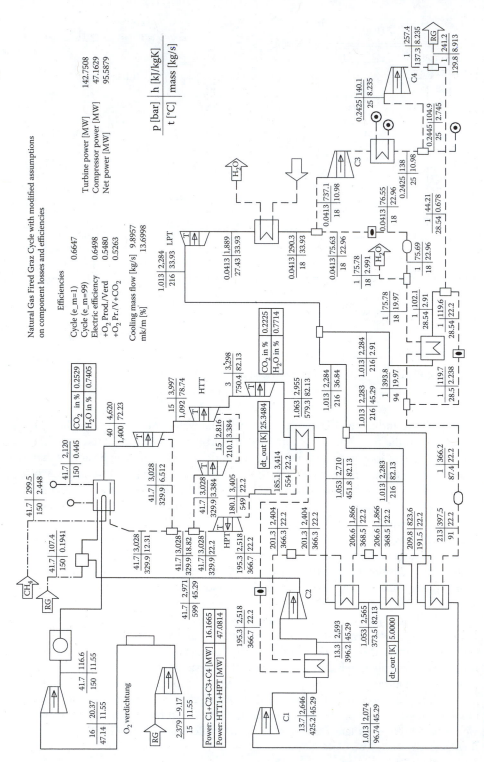

FIGURE 2.12 The current Graz cycle (Sanz et al., 2005).

Previously, academic calculations had been the basis of ZEPP research, but recently the industry has begun research in the area. The Zero Emissions Gas consortium, including some industry partners, are investigating combining a hydrogen production process and SOFCs to simultaneously produce electricity and hydrogen from natural gas with integrated CO_2 capture (Tomski, 2003). A similar project, also developed in the U.S.A., is FutureGen, an IGCC with precombustion capture and the use of hydrogen as a fuel, and CO_2 captured for enhanced oil recovery.

Aker Maritime began a long-term development of a commercial ZEPP in 1997. They are currently working with Alstom Power and other industrial partners on the development of a 25 MW plant for installation in the North Sea. The process produces separate streams of pure water and CO_2, using flue gas recycle. They mentioned that not only the CO_2 but also the N_2 produced by air separation may be useful for enhanced oil recovery.

ZECA (Zero Emissions Coal Alliance) is a group of companies that is developing a technology conceived at the Los Alamos National Laboratory that gasifies coal using steam in a process called *hydrogasification,* which produces pure hydrogen and pure liquid CO_2 for sequestration. This hydrogen fuels the gasifier and either an SOFC or a turbine, the exhaust of which provides the steam required by the hydrogasifier; see Figure 2.13. This new technology is not only zero emissions, but it also has double the efficiency of standard coal burning power plants. The technology can also be

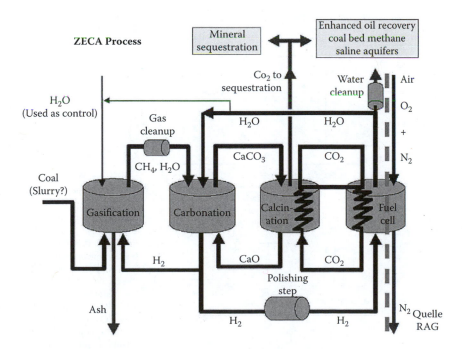

FIGURE 2.13 The schematics of Zero Emissions Coal Alliance (ZECA) process. (*Source:* www.cslforum.org/documents/TSRAppendix2003.pdf. Accessed Dec. 9, 2005).

adapted for other hydrocarbon fuels including biomass, and may have applications outside of power production, e.g., in oil refining processes (Tomski, 2003).

2.5 ZEPP CYCLES INCORPORATING OXYGEN ION TRANSPORT MEMBRANES

As previously mentioned, cryogenic air separation has a very detrimental effect on the performance and efficiency of ZEPP cycles. Oxygen ion transport membranes (OITMs) offer the possibility of oxygen production without a significantly adverse effect on the efficiency, and probably at a lower cost than cryogenic production. In fact, by using the air stream as a bottoming air turbine cycle, production of oxygen using OITMs can actually increase the efficiency of a ZEPP. OITMs currently have a maximum operating temperature of about 1000°C, and this can limit some cycles. The simplest ZEPP cycle incorporating an ion transport membrane (ITM) reactor is shown in Figure 2.14.

The ITM provides oxygen for combustion, with the heat of combustion used to provide heat to an ordinary Rankine cycle. The cycle is limited by the temperature of a Rankine cycle (state-of-the-art steam turbines have a maximum temperature of about 600°C), but this temperature is acceptable for the OITM ceramic. At a temperature of 540°C with a reheated Rankine cycle, the efficiency is 35.7% (Levin et al., 2003).

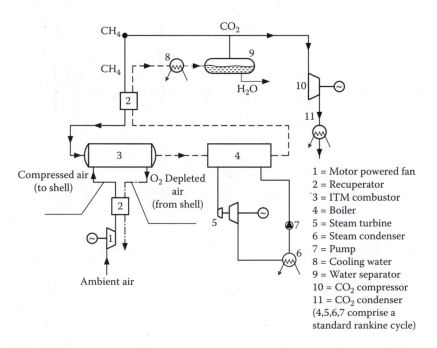

FIGURE 2.14 Simplest ZEPP cycle incorporating an ITM reactor (Levin et al., 2003).

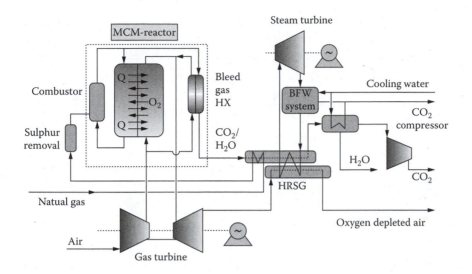

FIGURE 2.15 The AZEP cycle (Sundkvist and Eklund, 2004). MCM = mixed conducting membrane; HX = heat exchanger; BFW = boiler feed water; HRSG = heat recovery steam generator.

A well-developed example of a ZEPP incorporating ITMs is the AZEP (Advanced Zero Emission Power) cycle which combines a novel combustor integrated with a ceramic membrane and a heat exchanger (Sundkvist et al., 2001), shown in Figure 2.15.

Here, the simplified CO_2 portion of the cycle does not allow a high efficiency to be achieved. The intention of the AZEP authors is to avoid the use of CO_2 turbines, which are not currently available. They have integrated the OITM and combustion chamber: the membrane wall simultaneously conducts oxygen to the fuel side and heat to the air side. The main turbine is the air turbine (referred to in Figure 2.15 as the gas turbine), powered by the oxygen-depleted air, from which half of the oxygen has been removed. This turbine drives both the air compressor and the electrical generator. The combustion gases do not drive a turbine, and are used to provide heat to a bottoming Rankine cycle.

This cycle demonstrates the problem of using OITMs to provide oxygen: the OITM cannot be heated to too high a temperature, but the combustion should occur at the highest possible temperature for high efficiency using a gas turbine. The AZEP cycle was compared to a V94.3A combined cycle power plant, the efficiency of which is 57.9%.

> The penalty in thermal efficiency for the AZEP ... is 8.3 percentage points. This high loss is mainly due to the reduced turbine inlet temperature (1200°C) that causes significant power loss both in the gas turbine and in the steam cycle.

The turbine inlet temperature can be increased by optional firing of additional fuel in the heated air stream before entry to the gas turbine. The combustion products of

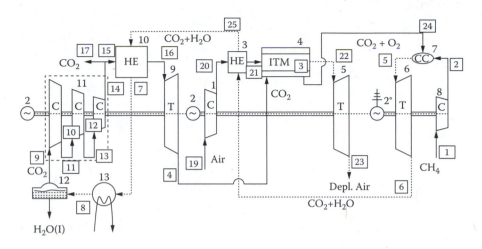

FIGURE 2.16 The ZEITMOP cycle (Yantovsky et al., 2003a). 1-air compressor; 2-synchronous electrical machine; 3-heat exchanger; 4-ITM reactor; 5-depleted air turbine; 6-CO_2 and H_2O turbine; 7-combustion chamber; 8-fuel gas compressor; 9-CO_2 turbine; 10-recuperator; 11-CO_2 compressor; 12-water separator; 13-cooling tower. (Numbers in boxes are node points.)

this additional firing are released into the atmosphere. By adding enough extra fuel at this point, the AZEP's efficiency is claimed to increase from 49.6 to 53.4%, but in this case, only 85% of the carbon dioxide is captured (Sundkvist et al., 2004). This retention rate is similar to that of a cycle with post-combustion CO_2 absorption.

An economic analysis of the AZEP cycle showed that a carbon emission tax of €31–€40/ton would make the AZEP with 100% carbon capture as economically attractive as the V94.3A plant (Sundkvist and Eklund, 2004).

If a ZEPP cycle is to be competitive with existing cycles without tax incentives, it must have a similar efficiency. This requires removing the restriction on cycle upper temperature caused by the membrane reactor. The ZEITMOP (zero emissions ion transport membrane oxygen power) cycle (Yantovsky et al., 2002) was developed independently of the AZEP cycle. The simplest version of this cycle is the gas-fired one, shown in Figure 2.16 and Figure 2.17, although the concept behind the cycle can also be applied to other fuels, e.g., pulverized coal.

In the ZEITMOP cycle, the ITM reactor is remote from the combustion chamber, allowing much higher combustion temperatures to be achieved. After separation of combustion products, the carbon dioxide is cooled and compressed, then heated and expanded (i.e., a Rankine cycle) before entering the ITM to be mixed with oxygen. This mixture then enters a separate combustion chamber. As a result, the ZEITMOP cycle can have a higher combustion temperature and a higher efficiency. If the turbine inlet temperature is 1500°C, the ZEITMOP cycle efficiency is claimed as 56%. This temperature limit depends only on the turbine, not on the ITM reactor. Current turbine inlet temperatures are on the order of 1300°C, at which temperature the ZEITMOP efficiency is claimed as 46%.

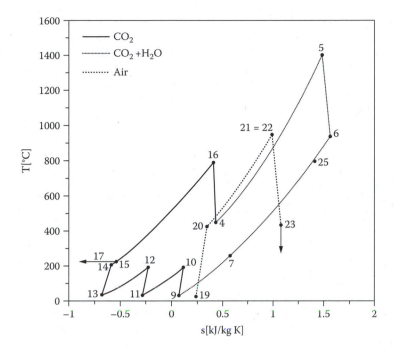

FIGURE 2.17 T–s diagram of ZEITMOP cycle (Yantovsky et al., 2003a). (See also Table 2.4)

The ZEITMOP cycle has not been globally optimized, so there remains hope for an increase in efficiency through optimization. As this cycle could be developed for all types of fossil fuels, it represents one of the best options for the replacement of decommissioned power plants.

TABLE 2.4
Data for ZEITMOP Cycle

Basic Data for ZEITMOP Cycle—Calculation of Energy Balances for Turbine and Compressor Units

$N9 = N_{HPT} = m_4 (h_{16} - h_4) = 14.08$ MW	CO_2 high-pressure turbine (element "9")
$N5 = N_{AirT} = m_{23} (h_{22} - h_{23}) = 9.10$ MW	Depleted air (Dair) turbine (element "5")
$N6 = N_{GasT} = m_6 (h_5 - h_6) = 27.91$ MW	CO_2+H_2O (Gas) turbine (element "6")
$N11 = N_{CO2} = m_9 [(h_9 - h_{10}) + (h_{11} - h_{12}) + (h_{13} - h_{14})]$ $= 16.98$ MW	CO_2 compressors (element "11")
$N1 = N_{Air} = m_{19} (h_{19} - h_{20}) = 8.0$ MW	Air compressor (element "1")
$N8 = N_{Fuel} = m_1 (h_1 - h_2) = 0.65$ MW	Fuel (CH_4) compressor (element "8")

Note: Net turbine power: $N_{net} = [(N9 + N5 + N6) - (N11 + N1 + N8)] \times \eta_m = 25.2$ MW, (at the mechanical efficiency: $\eta_m = 0.99$). Thermal efficiency of principal ZEITMOP cycle: $\eta_{th} = N_{net}/(m_1 Q_d) = 0.5038 = 50.38\%$ (at lower heating value for CH_4: $Q_d = 50$ MJ/kg).

Source: Yantovsky, E. et al., 2003a.

The use of CO_2 turbines should not be an insurmountable technical problem. Such turbines were investigated by the Esher Wyss Co. in Switzerland about 40 years ago (Keller and Strub, 1968). McDonnell Douglas in the U.S.A. built and tested a microturbine unit using supercritical CO_2, which produced 150 kW at an efficiency of 29% (Hoffman and Feher, 1971). The temperature and pressure at the turbine inlet were 1005 K and 22.95 MPa, with a flow rate of 2.75 kg/s.

Mathieu (1994) carried out a simple model of a CO_2 turbine, which showed that due to differences in molecular weight and adiabatic expansion coefficients, air-based turbines need to be completely redesigned to operate with CO_2. According to similarity laws, a reduction of the rotational speed of an air-based GT should accommodate an operation on CO_2 instead of air. However, the properties of CO_2 vary much more with temperature than those of air so that a full redesign of the machine is unavoidable.

Work carried out on the Graz cycle has significantly advanced the development of CO_2 turbines (Jericha et al., 2003; Heitmeir et al., 2003).

Siemens, Westinghouse, and Praxair collaborated to develop a zero emissions fuel cell cycle (Shockling et al., 2001). In a hydrocarbon-fueled fuel cell, the fuel is fed to the anode side of a fuel cell and air to the cathode side. About 85% of the fuel is used in the fuel cell, and the gas leaving the anode side is normally mixed with the cathode gas and they are burned together. The heat from combustion is used to preheat the incoming air and fuel, and also to partially reform the fuel. In the system described here, the cathode gas is fed to one side of an oxygen ion transport membrane, with the anode (air) gas on the other side. Oxygen passes through the membrane to completely oxidize the cathode gas stream, which then consists entirely of carbon dioxide and water vapor. Fuel cell cycles are very efficient, so this cycle is a promising development in the area of zero emissions cycles.

The membrane reactor in this cycle is being developed by Praxair, who is focusing on a tubular membrane concept. The paper gives the results of many tests on the reactor.

Another coal-fired ZEPP incorporating ITMs is the Milano cycle (Romano et al., 2005). It is similar to the AZEP cycle due to the lack of CO_2 turbines and restriction on cycle temperature due to the membrane reactor. The bottoming cycle is an ordinary steam cycle, which generates power, whereas the air turbine drives the compressor only. The cycle is shown in Figure 2.18, along with some technical data, given in Table 2.5.

Increasing the combustion pressure from 1.15 bar up to 10 bars resulted in a negligible increase in the calculated efficiency, which is quite standard for a Rankine cycle (41.34–41.89%). Solving the membrane reactor problem is required to make this cycle competitive with other coal-fired ZEPPs.

The Oxycoal-AC cycle, presented by Renz et al. (2004, 2005), includes a high temperature membrane unit, in which oxygen is mixed with carbon dioxide and water vapor. Pulverized coal is burned in this mixture to provide heat for a Rankine cycle. The system is shown in Figure 2.19.

Note the depleted air is described as N_2 in the diagram. This is incorrect as only oxygen has been removed from the air. Other elements remain, e.g., water vapor, argon, carbon dioxide. Also, it is impossible for 100% of the oxygen to be removed by the membrane. Some must remain as the oxygen partial pressure on the feed side must be greater than that on the permeate side. The efficiency for this cycle, 41%,

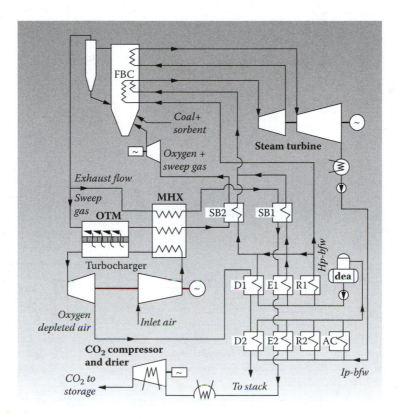

FIGURE 2.18 The Milano cycle (Romano et al., 2005).

TABLE 2.5
Performance of the Milano Cycle

FBC-USB Performance Calculation

FBC pressure, bar	1.15	10
Coal LHV input, MW	984.9	984.9
Turbocharger inlet air flow, kg/s	593.5	698.2
Turbocharger pressure ratio	11	20
Turbocharger power output, MW	74.7	69.7
ST power output, MW	404.9	400.6
Net power output, MW	407.2	414.6
Net plant LHV efficiency, %	41.34	41.89

Source: Romano, M. et al., 2005.

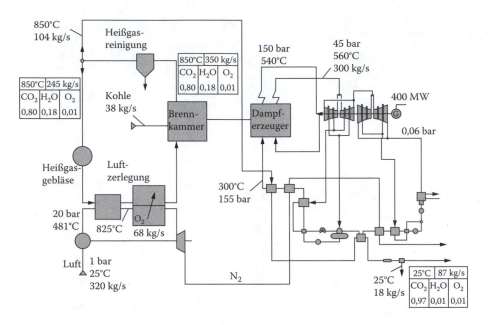

FIGURE 2.19 The Oxycoal-AC cycle (Renz et al., 2004). Kohle = coal; Luft = air; Brennkammer = combustion chamber; Dampferzeuger = steam generator; Heißgasreinigung = flue gas cleaner; Heißgasgebläse = flue gas pump; Luftzerlegung = air separator (www. oxycoal.de/index.php?id=336).

was calculated based on a simulation using Ebsilon (Renz et al., 2005). It is very similar to the efficiency of the Milano cycle, which is unsurprising as the two cycles have no major differences.

Commercial interest in oxycoal is strong. The U.S. Department of Energy has granted funding to Babcock and Wilcox, who are using oxy-combustion of coal in wall-fired and cyclone boilers (Anna, 2005). Vattenfall plans to build a 30 MW_{th} pilot plant using oxyfired combustion of coal in Germany (the Schwarze Pumpe lignite fired plant). This is scheduled to begin in 2008; see Figure 2.20.

Further development of coal-fired plants is presented in the coal gasification and integrated carbon dioxide capture, described in the German program of emission reduction referred to as COORETEC 2003. These schematics (see Figure 2.20) are very promising for coal-rich countries, especially China and India.

2.6 ZERO EMISSIONS VEHICLE CYCLE—PRELIMINARY SECTION (ALSO SEE CHAPTER 7)

Carbon dioxide emissions from vehicles are more of a problem than those from power plants. Vehicle manufacturers seem to focus entirely on hydrogen or electric vehicles in their attempts to create a zero emissions vehicle. Unless the hydrogen or electricity is produced by a zero emissions process, these vehicles are not zero emissions. The only truly zero emissions vehicle cycle is the Zero Emissions Membrane

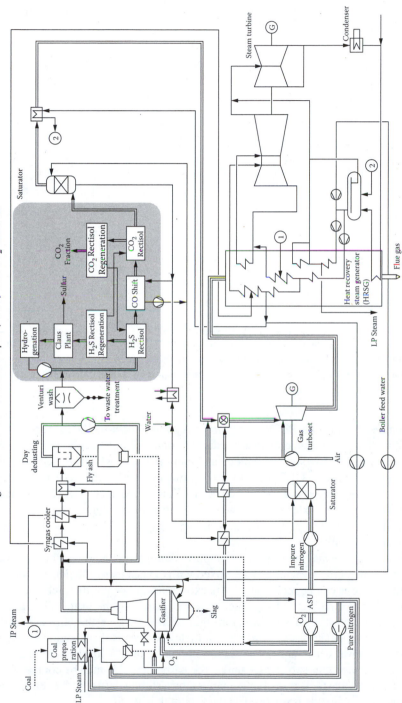

Integrated Gasification Combined Cycle (IGCC) with CO₂ Removal

With available fuels, proven technology and proven components, the cycle with CO_2 removed does have efficiency of 42–45%.

FIGURE 2.20 The coal-fired ZEPP developed in the German program COORETEC. (*Source:* Essen University, http://www.cooretec.de/; accessed Sept. 11, 2007.)

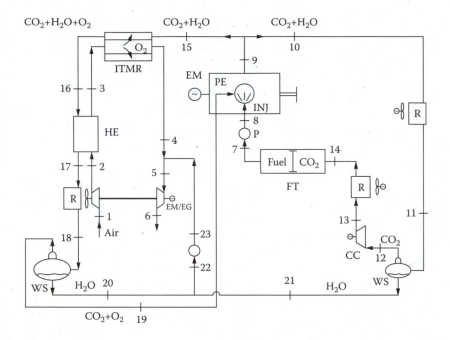

FIGURE 2.21 The ZEMPES cycle (Yantovsky et al., 2004). CC = CO_2 compressor, EG = electric generator, EM = electric motor (for starting), FT = fuel/CO_2 tank with sliding baffle, HE = heat exchanger, INJ = fuel injection, ITMR = ion transport membrane reactor, P = pump, PE = piston engine, R = radiator-cooler, TC = air turbocompressor, WS = water separator.

Piston Engine System (ZEMPES) cycle (Yantovsky and Shokotov, 2003b; Yantovsky et al., 2004, 2005). Cryogenic oxygen production onboard a vehicle is likely to be unfeasible due to the unavoidable vibration and inertial forces, which are detrimental to distillation columns, so OITMs are the most attractive option for oxygen production onboard a vehicle. The ZEMPES cycle uses an OITM reactor (ITMR in the schematic) to oxygenate exhaust gases, which are recirculated to the ordinary piston engine. The simplest version of the cycle is shown in Figure 2.21, along with technical information in Table 2.6.

The piston engine may be a spark or compression ignition engine. The exhaust gases heat the membrane reactor and are mixed with oxygen there. Obviously, not all of the exhaust gases can be recirculated; the extra portion born of the fuel and oxygen combustion is removed and separated by condensing out the water. This water may be injected into the air stream before the turbine, making the only emission harmless water vapor (just as in a hydrogen vehicle). Alternatively, it could be stored onboard for removal when the vehicle is refuelled, as is done with the carbon dioxide. To avoid the need for two storage tanks, the carbon dioxide may be compressed and stored in the same tank as the fuel, separated by a sliding baffle. At the refuelling station, the carbon dioxide is removed as the fuel tank is filled.

TABLE 2.6
ZEMPES Efficiency

Parameter	Units	Results
Turbine power	kW	106.35
Air compressor power	kW	93.16
CO_2 compressor power	kW	34.23
Total power	kW	127.4
Piston engine indicator power	kW	283.6
Piston engine effective power	kW	235.5
Friction losses	kW	11.76
Radiators fan power	kW	26.7
Fuel energy input	kW	800
Fuel consumption	kg/hour	64.2
Specific fuel consumption	g/kWh	272.4
System efficiency	%	28

Source: Yantovsky, E. et al., 2004.

The calculated efficiency of 28% seems to be acceptable for a zero emissions vehicle. It can be increased to 37% by addition of an exhaust gas turbine and to 44% by addition of a bottoming Rankine cycle (Yantovsky et al., 2005). However, such a complicated system is too cumbersome for a real vehicle. Reducing the dilution of the combustion mix may provide an easier method of increasing the efficiency.

Startup of ZEMPES can be easily implemented by switching off the recirculation, burning the fuel in air, and using the exhaust gases for heating the membrane only. This requires allowing some emissions at startup. Measuring the carbon dioxide stored onboard allows easy tracking of these emissions. Alternatively, some oxygen could be stored onboard for zero emissions combustion during startup.

2.7 TOWARD A ZERO EMISSIONS INDUSTRY

All zero emissions cycles might be considered as cogeneration of power and carbon dioxide. The quantity of carbon dioxide so produced will probably exceed the industrial demand. The greatest consumer of carbon dioxide is EOR (Enhanced Oil Recovery) and ECBM (Enhanced Coal Bed Methane Recovery). The contemporary and future convergence of the power industry and the oil and gas industry on a zero emissions basis was considered by Yantovsky and Kushnirov (2000b) and some relevant cycles were discussed by Yantovsky (2000c).

Akinfiev et al. (2005) have shown the ultimate goal of CO_2 injection underground: a possible method of converting CO_2 to methane through reaction with fayalite.

The worldwide capacity of gas-fired power plants suitable for AZEP technology from 2020 was estimated to be in the range of tens of GW. Later on, it is estimated that the AZEP cycle could be commercially available in less than 10 years, given a market (Sundkvist and Eklund, 2004).

As mentioned throughout this book, the Clean Energy Systems plant will soon be joined by a number of other demonstration and commercial plants. In general, it appears that ZEPPs are close to commercialization. The interested reader may find excellent reviews of ZEPPs in Göttlicher (1999, 2003), Bolland (2004a, 2004b), Gupt (2003) and Bredeson et al. (2004). Only the last reference contains ZEPPs with oxygen ion transport reactors. A recent comprehensive review of carbon capture technologies is given by twelve leading professionals from ten power companies in VGB (2005). Unfortunately, this review almost ignores ITM reactors for oxygen production, and mentions only a few membrane technologies, including the AZEP cycle, as examples of futuristic technologies. The authors of this report claim that carbon capture always reduces electrical efficiency of a cycle. However, this is not necessarily true. For example, in the ZEITMOP cycle, the recirculated carbon dioxide undergoes a Rankine cycle, actually adding to the electrical efficiency.

Most advanced seems to be the project of Germany's the brown-coal-fired ZEPP or Schwarze Pumpe. The general outlook can be seen in Figure 2.22.

Quite naturally, the selection of the best power cycle depends on economics, namely on cost of energy (COE). At the end of the book, we give a general optimization method (Pareto optimization) which includes COE. In Figure 2.23, we present some preliminary figures on the comparison of different zero emissions power plants. In general, the difference between COE of cycles of Zepp and not Zepp is modest, within the accuracy of calculations.

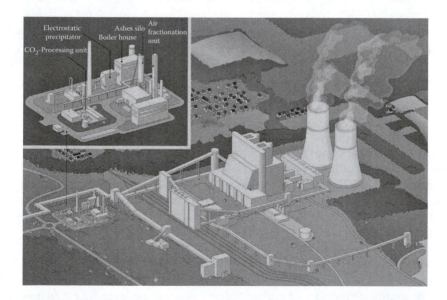

FIGURE 2.22 Outlook of Schwarze Pumpe ZEPP. It should be started in 2008. (*Source:* http://www.vattenfall.com/; accessed Nov. 11, 2007.)

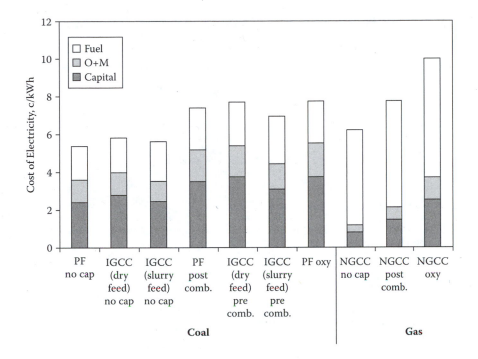

FIGURE 2.23 Comparison of COE of different ZEPP. (*Source:* Report IEA GHG R&D Programme, May 2007.)

2.8 AN IMPORTANT PAPER

After reviewing almost all ZEPP thermodynamic cycles, let us take a look at Figure 2.24, which is a photocopy of a paper from 1991, containing some schematics and wordings that remain of interest today.

For example, in the recent paper by C. Salvador of CANMET, presented at the 2nd Oxy-combustion Workshop (2007) new wording, *Hydroxy-fuel Technology,* is introduced where water is used to moderate combustion of natural gas in oxygen.

Here, *new* is only new wording. The schematics do not differ from the CES cycle and GOOSTWEG cycle (1991) seen in Figure 2.24.

2.9 SOME ADDITIONAL REMARKS

On January 24–26, 2007 in Windsor, Connecticut, U.S.A., an important event took place, the 2nd International Oxy-Combustion Workshop organized by IEA GHG and sponsored by ALSTOM Power Inc. It was attended by over 85 people from 16 different countries. There were many papers presented on very advanced works, including encouraging data on the ITM reactors created in Air Products and Praxair.

It is interesting to note that 10 years before, at Liege University, on exactly the same date—January 24, 1997— a similar workshop occurred.

<u>World Clean Energy Conference (WCEC), 4 - 7 November, 1991 in Geneva</u>

THE THERMODYNAMICS OF FUEL-FIRED POWER PLANTS
WITHOUT EXHAUST GASES

E. I. Yantovskii
Institute for Energy Research
U.S.S.R. Academy of Sciences
117333 Moscow, U.S.S.R.

ABSTRACT

An attempt to find a radical solution to the problem of atmospheric pollution, especially to prevent the release of greenhouse gases, is presented Several power plants along with their thermodynamic cycles on a t-S diagram are discussed. In each, the nitrogen is eliminated before combustion by means of air splitting. Furthermore, all the exhausts are liquified and are disposed of in the depths of the sea or are used to neutralize alkaline ash ponds.

The analysis shows that in spite of a significant consumption of power to get oxygen and compress carbon dioxide, the efficiency may be the same as for ordinary cycles.

INTRODUCTION

In ancient times, the contents of night pots in some European countries was thrown out of windows into the streets. After this practice was forbidden, some citizens used to stealthily at night do the same over the fences of their neighbors. The very high chimneys of our fuel-fired power plants are in effect devices for disposing of waste gases and particulate matter to our neighbors who tolerate this and behave likewise.

However, recently it was recognized that as a result there is a greenhouse effect and all the neighbors have become the inhabitants of one apartment. The greenhouse effect is a global problem indeed; and unfortunately, great investments are needed now in order to protect the future. Till now little progress has been made, in spite of strict environmental regulations.

Looking at power plants, a real start might be made if we recognize not only what the problem is, but also what the solution should be. We must find for future power plants not only the cycle which is economically attractive, providing substantial energy conservation; but one which is also ecologically benign.

The aim of this paper is a preliminary assessment of several schemes and their thermodynamic cycles. At this early stage, we restrict ourselves to the thermodynamics involved, without any economic assessment. The latter must, of course, be done at some later point by means of a computer exergonomics approach [1].

The schemes which are considered are[1]

- the GOOSTWEG power cycle: gas or oil fired oxygen steam turbine without exhaust gases - recommended for gaseous or liquid fuels

------

[1] An additional type of thermodynamic cycle for liquid gas fuelled cars is also presented at the end of the paper. It is called the Automotive Steam Turbine Without Exhaust Gases (AMSTWEG) cycle .

FIGURE 2.24 Photocopy of a 1991 paper. (*Source:* The Thermodynamics of Fuel-Fired Power without Exhaust Gases. Paper presented at the World Clean Energy Conference, Nov. 4-7, 1991, Geneva.)

- COMHDWEG power cycle: coal fired oxygen magnetohydrodynamic generator without exhaust gases - recommended for coal or low grade fuels such as shale oil

- COGSTWEG power cycle: coal-oxygen gasified steam turbine without exhaust gases - recommended for coal and shale oil use

- SOFT power cycle: solar oxygen fuel turbine for solar energy use.

All these cycles serve one primary idea which is to eliminate the release of harmful exhaust gases by means of the elimination of nitrogen before combustion. Certainly the simplest case is with the use of natural gas. The author believes that there is at least more than a hundred years supply of this fuel left in the world in view of the off-shore and hydrate resources which are available. Many countries such as the Soviet Union, the United States, the United Kingdom, the Netherlands and others will use it for a long time.

The coal option is also important. However, in this case pollution which is the main problem must be strictly prevented.

The last but not least type of fuel, solar energy, is the ultimate fuel for the development of power production. Here the SOFT cycle seems to be extremely useful. According to Professor John Bockris,

(t)he major difficulty is that the efficiency of light conversion to energy via biomass is small (about 1%), and most countries do not have sufficient free land-area on which to accommodate the necessary trees or other crops. If one could use genetic manipulation to make biomass grow at a 10% per year energy-conversion efficiency, then the possibility of biomass providing a solution would be greatly advanced [2].

The SOFT cycle involves well developed photosynthesis with just 10% efficiency by means of ponds with microalgae.

THE GOOSTWEG CYCLE

The plant outline is presented in Figure 1. Air enters splitting machine 1 where it is divided into nitrogen and oxygen. The former is released to the atmosphere without any harm or is used elsewhere. Oxygen mixed with high pressure feedwater in 2 enters combustion chamber 3 to which compressed methane is directed. Steam injection into the combustion chambers of the gas turbines is widely used, for example, by General Electric in its STIG cycles [3]. Our case differs due to the almost complete elimination of nitrogen before combustion.

The temperature prior to the turbines is 750 °C and is enough to get acceptable efficiencies without special blade cooling. Possible increases in temperature up to those used in modern gas turbines (up to 1200 °C) are a chance for future improvements.

After triple heating and expansion in turbines 4, 6 and 8, the mixture of steam and CO_2 enters recuperator 9, heating the feedwater, fuel and oxygen in separate heat exchangers. The steam is then condensed in the condenser and returned. The carbon dioxide from the degaser enters the compressor and in liquid form is disposed.

In all the calculations, the well known M.P. Wukalovich Tables of the thermodynamic properties of water and steam were used (seventh edition, Moscow, Energy, 1965), with some corrections for the addition of CO_2. Note, however, that there is some degree of uncertainty in these corrections.

FIGURE 2.24 (Continued).

The cycle is presented on Fig. 2 in t-S coordinates. Its principle state points are as follow:

Table 1 - Thermodynamic state points for the GOOSTWEG cycle.

Point	T (°C)	P (bars)	s (kJ/kg K)	i (kJ/kg)
a	80.	.5	1.1	.33
b		240.	1.1	.39
c	200.	240.	2.3	.86
d	750.	240.	6.8	3.91
e	420.	30.		3.25
f	750.	30.	7.9	4.02
g	460.	5.		3.38
h	750.	5.	8.7	4.04
i	390.	.5		3.25
k	80.	.5	7.6	2.64

The isentropic efficiency of the turbines is assumed to be 0.90. The equilibrium reaction of combustion is as follows:

$$CH_4 + 2O_2 \rightarrow CO_2 + 2H_2O + (55 \text{ MJ/kg } CH_4)$$
$$(16) \quad (64) \quad\quad (44) \quad (36)$$

The fuel-air ratio based on the stoichiometric coefficients is 64/16 = 4, whereas the enthalpy rise during combustion per kg of combustion products is 55/(1+4) = 11 MJ/kg. The enthalpy rise after combustion per kg of the steam-CO_2 mixture is

cd	3.91 - 0.86 =	3.05 MJ/kg
ef	4.02 - 3.25 =	0.77
gh	4.04 - 3.38 =	0.66
total		4.48

The required relation of the total flow rate to combustion products is 11/4.48 = 2.46. According to molecular mass of carbon dioxide, its flow rate share is (44/80)(2.46) = 0.224. For CH_4 it is 0.081 and for O_2 0.325.

In many textbooks, the specific power consumption to produce oxygen is around 0.3 kW.h/kg O_2. Perhaps in some new devices where absorption or membrane technology is used, it may be less, but for the sake of reliability, this figure shall be used. Thus, per kg of the working fluid, it is

$$D_{O_2} = (0.3)(3.6)(0.325) = 350 \text{ kJ/kg}$$

CO_2 compression from subatmospheric pressure up to some tens of bars is needed to cross the saturation line. In Fig. 3, the process is presented as a straight horizontal line but in fact many "teeth" exist. Power consumption is approximately evaluated by the product of the temperature, entropy change and mass share divided by the compression efficiency, i.e.,

$$D_{CO_2} = (0.224)(288)(5.95-4.95)/0.7 = 92 \text{ kJ/kg}$$

The cycle efficiency (see the state points in Fig. 2) is

FIGURE 2.24 (Continued).

$$\frac{de + fg + hi - D_{O_2} - D_{CO_2}}{cd + ef + gh} = \frac{660 + 640 + 790 - 350 - 92}{4480} = 0.37$$

This figure is comparable to that of ordinary plants with stacks.

The numerator above, 1648 kJ/kg, is the specific power per kg of flow. Hence, for a power plant of 1.6 GW, a flow rate of 1000 kg/s is needed. In this case, the CO_2 flow equals 224 kg/s. Assuming a velocity of liquid CO_2 of only 1 m/s and a density of 1250 kg/m^3, the required pipe cross-section is 224/1250(1) = 0.18 m^2. The pipe diameter is 0.48 m.

Looking at this figure we may conclude that disposing of the liquid carbon dioxide even with long distances is a mild problem. It could be used to neutralize some basins containing alkalies. The best global storage of CO_2 certainly might be the ocean, as is clear in Fig. 4. The use of carbon dioxide in the ocean to enhance photosynthesis is worth attention. The absorption capacity of the ocean many times exceeds the ability of mankind to burn fuel. Even though the solubility of CO_2 in low pressure warm water is quite small, in deep cold water it is very high (see Fig. 5 [4]).

DISCUSSION OF THE GOOSTWEG CYCLE

This cycle is not new. Many of its important features have appeared in earlier papers. One of these is by Kuzminsky and another by Christianovich [5]. From 1892-1900, a combustion chamber with steam injection was developed by P.D. Kuzminsky. In his test rig, the steam of high pressure water flowing in the wall of the chamber was used. In the early sixties, a group headed by S. A. Christianovich developed a power process which our cycle resembles closely. The steam injection into the combustion gases, three combustion chambers and turbines for high, medium and low pressures and a maximum temperature of 750 °C are the same. Their experimental work on combustion chambers and other components is a source of useful information. Unfortunately, after initial success, these investigations were cancelled.

Two features set the GOOSTWEG cycle apart from that developed by Christianovich. They are the complete elimination of nitrogen and the disposal of the liquefied effluents in the deeps, features which are necessary in view of the *greenhouse effect* which at the time of Christianovich was unknown.

The GOOSTWEG cycle also requires the use of a big air splitting machine which for a 1 Gw unit is an order of magnitude greater than any existing on the market today. However, this technology is well developed so that the required equipment can be constructed.

THE COMHDWEG CYCLE

The problem of the ecologically benign use of coal and low grade fuels for power production will last for a long time. An attempt to solve this problem in the wake of the previous approach is presented here.

In view of some of the particular properties of this cycle, we must restrict ourselves to power plants near a sea coast. There are quite a few of such plants in Europe, Japan, Israel and other countries.

The scheme of this cycle with its proposed MHD power plant is presented in Fig. 6. Air from the atmosphere enters the splitting plant where it is separated into oxygen and nitrogen. The latter is released back into the atmosphere without any chemical

FIGURE 2.24 (Continued).

reactions and is, therefore, harmless. The pure oxygen which is compressed up to 100 bars via a stainless steel heat exchanger enters the combustion chamber.

The fuel consists of coal powder particles mixed in the strong brine of sea water, the waste of desalination. In effect, it is the slurry of fuel and seed (i.e., the potassium chloride of sea salt). The combustion of this slurry in pure oxygen preheated up to 1100 $^\circ$K creates a very high temperature of nearly 3900 $^\circ$K.

At the exit of the combustion chamber, some part of the liquid slag is released while the other part continues to flow in the MHD channel to protect the walls and electrodes. After isentropic expansion in a nozzle, the combustion gases with their high conductivity enter the MHD channel and the conducting gas (plasma) flowing in a transverse magnetic field creates an electromotive force and electrical current in the circuit which consists of electrodes, gas stream and external load resistance [6-10]. It should be noted that in view of the very large pressure ratio required (100 to 1), a disk shaped MHD generator with circular magnetic lines might be recommended (see Fig. 7 [10]).

After the MHD channel, the residual slag has to be absorbed along with the rest of the seed while hot carbon dioxide via a diffuser enters the boiler of an ordinary steam cycle. Some of its heat is used to preheat the oxygen. It is assumed that all the seed was absorbed by the liquid slag and after cooling is consumed in the construction industry or elsewhere. No seed return is needed since for coastal plants the strong brine in the fuel is not worth saving.

In the boiler, the flow of carbon dioxide is cooled down to the sea water temperature at which point the primary new process of the cycle begins, i.e., the liquefaction of the carbon dioxide by means of a simple isothermal compression. In fact there are no pure isotherms in real processes so that such processes actually consist of many adiabatic compressions and intercooling by means of sea water. A temperature of about 15 $^\circ$C was selected (Fig. 3).

Compressed wet CO_2 vapor is directed through a steel pipe to the depths of the sea (500 m) where it can be released without damage to the atmosphere. The world's oceans can absorb all the man-made CO_2. In the global carbon cycle, the oceans play the most important role of dissolved storage as is evident in the diagram of Figure 4 from B. Giovannini and D. Pain [11]. Note here that the dissolved inorganic (36700 Gt) exceeds 3-6 times the total mass of carbon fuels and 50 times the amount of carbon in the atmosphere (725 Gt).

NUMERICAL RESULTS FOR THE COMHDWEG CYCLE

The I-S diagram and t-S diagram for the proposed cycle are presented in Refs. [12] and [13], respectively. It is assumed that the oxygen enters the power plant with a pressure of 100 bars and a temperature of 20 $^\circ$C. Its enthalpy $I = 0$. At point one, the oxygen receives $I = 1$ MJ/kg from a heat exchanger.

According to the diagrams in the above references, the combustion heat for Donets-basin coal is $I_r = 32$ MJ/kg. The stoichiometric ratio is $L = 2.6$ in kg O_2 per kg of C (Fig. 8). The enthalpy rise per 1 kg of mixture is

$$I_r / (1 + 2.6) = 32 / 3.6 = 8.9 \text{ MJ/kg } CO_2$$

and the heat loss in the combustion chamber (5%) = .45 MJ/kg CO_2

All the state points of this cycle are presented in Fig. 8 and in Table 2. As in any preliminary thermodynamic cycle estimations, some reasonable assumptions on the efficiencies of the main processes are needed. They appear in Table 3.

FIGURE 2.24 (Continued).

Table 2 - Thermodynamic state points for the GOMHDWEG cycle.

Point	I (MJ/kg)	t ($^{\circ}$K)	P (bars)	Content
0	0.0	288.0	100.0	O_2
1	1.0	1100.0	100.0	O_2
2	9.45	3900.0	100.0	CO_2+seed +dissociate
3	5.4	2740.0	1.0	CO_2 gas
4	4.3	2540.0	1.0	CO_2 gas
5	0.0	288.0	1.0	CO_2 vapor
6	0.0	288.0	50.0	CO_2 liquid

Table 3 - Process efficiencies for the GOMHDWEG cycle.

	(m)	(w)	(b)
combustion chamber	0.95		
MHD isentropic including excitation	0.7	0.6	0.8
steam bottoming	0.4	0.35	0.45
isothermal CO2 compressor	0.7	0.6	0.8
calculated cycle efficiency	0.385	0.31	0.46

In this table, the moderate (m), worst (w) and best (b) cases are presented. Even the worst case is suitable in view of the cleanliness of the plant.

The power consumption to get $1 nm^3$ of O_2 is assumed to be 0.4 kWh which translates into 1MJ/kg O_2 or .72 MJ/kg CO_2. The power to compress CO_2 is equal to $T_o(S_5-S_6)$/efficiency which equals 288(5.95-4.95)/0.7=.41 MJ/kg in the moderate case.

The efficiency calculation for this case is as follows:

cycle efficiency = [(4.05)(.7) + (4.3)(.4) - .41 - .72]/8.9 = .385

The numerator here is the specific power per 1 kg CO_2 and is equal to 3.12 MJ/kg. Hence, for a 1 GW plant, the gas flow rate is

$$\frac{1000 \text{ MJ/s}}{3.12 \text{MJ/kg}} = 320 \text{ kg/sec}$$

This rather small flow rate is evident because of the lack of nitrogen ballast. At the end of the isothermal compression to 15 °C and 50 bars and with a density of CO_2 equal to 143 kg/m^3, its velocity in a pipe having a 0.5 m diameter is quite reasonable, i.e., 13 m/s.

FIGURE 2.24 (Continued).

The liquid CO_2 velocity in the same pipe at its exit in the sea depths is 1.6 m/s. It is quite clear that transferring all the release from a single large 1 GW MHD unit by means of an ordinary steel pipe half a meter in diameter poses no problem.

THE COGSTWEG CYCLE

This cycle is based on coal gasification. There exist many such processes [14]. Almost all of them use steam and oxygen. For our purposes, the cleanest and most well developed process has to be chosen. Three examples of such processes are the Lurgi, Texaco and Bigas processes. In the Texaco process, the pressure is not more than 18 to 35 bars. Hence the produced fuel gas has to be compressed up to 240 bars. There are no exhaust gases which is very important for the power plant cycle.

The plant outline is presented in Fig. 9. The new elements are a coal gasifier (14) and ash ponds (11). The gas of 14 is much less caloric than methane, but this is of no importance because of the regulation of steam flow into the combustion chamber. In the particular case of alkaline ash, this plant might be very clean, providing the means of neutralizing the ash by means of the carbonic acid of dissolved CO_2.

If the gasifier efficiency equals 0.75, then the total power plant efficiency related to the coal input is $(.75)(.39) = 0.29$. This efficiency could obviously be increased by increasing the temperature at the turbine inlet; but even with the rather low efficiency of 0.29, this cycle could be attractive for coal consuming countries.

The COGSTWEG cycle offers a lot of possibilities for producing extremely clean, nature conserving power. Typical emissions in the U.S.A. for coal-fired plants per GJ of energy delivered is as follows: $CO_2 = 240$ kg, $SO_2 = 90$ g, $NO_x = 90$ g and suspended particles equal to 6 g. Among 19 projects of clean coal-fired power plants [18] including the Integrated Gasification Combined Cycle (IGCC), not a single radical solution has appeared. Current reports indicate that the removal of CO_2 from combustion gases would raise the cost of electricity by 35 to 80 % with the final rate of CO_2 still an uncertainty. The COGSTWEG cycle seems to be helpful here.

THE SOFT CYCLE

The layout for the SOFT power plant is presented in Fig. 10. Air enters the splitting machine (1) where it is divided into nitrogen and oxygen. The nitrogen is released to the air. In the mixer (2), oxygen flow is injected into hot high pressure steam. This mixture and the fuel enter the first combustion chamber 3 where the temperature of the steam and CO_2 as combustion products increases to a maximum.

Steam expands in the high pressure turbine (4) which drives an electrical generator. After the first expansion, the fuel and oxygen are injected and burned in chamber 5 and then again in chamber 7. After the low pressure turbine (8), the mixture of CO_2 and steam enters the recuperator (9) where partial condensation takes place. However, the CO_2 is still a vapor. The total condensation of steam and the dissolving of CO_2 occurs in the jet condenser (10) where pumped water from the pond (14) is used.

The pond (14) is the most important part of the plant. It is filled by fresh water with photosynthesis making algae chlorella or something like that. The aim of the pond is to absorb solar energy in order to convert dissolved CO_2 into organic matter which is then used as fuel for the power plant. The oxygen from the reduced CO_2 is released to the atmosphere. Under steady-state conditions, the production and consumption of oxygen are equal.

These *renewables* depend mainly upon the exergy flow concentration. As a rule, the natural exergy current density is low, on the order of 100W/m^2 (on the average).

FIGURE 2.24 (Continued).

However, some rare exceptions exist. One is the hot jets of *blacksmokers* in the ocean depths [1] where at a temperature of 350 °C the exergy current reaches 1Gw/m².

One of the best concentrators of natural exergy is photosynthesis. In the global energy flow diagram in Fig. 11 [15], one can see something resembling a law of natural energy flow, i.e., the less the flow the higher the concentration. Large flows with hot humid air (comprising 97.5 % of total solar energy input) have very low current densities such as the one mentioned above (0.1kW/m²). Warm water and ocean currents are nearly 3 % of the total solar energy input and have a current density of 1MW/m². If we consider the flow of water with microalgae, e.g., green river water, with the suspension of only 1 % of organic matter, a velocity of 2 m/s and a combustion heat of 50 MJ/kg, the current density is

$$10^3 \ (2)(50)(10^6)(10^{-2}) \ = \ 1GW/m^2$$

Hence, in the case of biomass which is only 0.04 % of the total solar energy input, the current density might be a million times greater than in moving air or in solar light, keeping in mind the solar constant of 1.3 kW/m².

The water treatment plant (12) also acts as a refinery, dividing the mixture of water and algae into pure high quality water and organic matter which is then used as fuel. The concentration of energy takes place at first in photosynthesis itself, then in the flow of water with organic material and finally in the plant itself (12) when water is separated from the fuel. Here a slurry of organic matter in water may be used.

The primary energy of this cycle is solar falling on the ponds surface. The converted energy is in the form of electricity from the generators. The losses are heat losses from equipment and low grade heat losses including evaporation from the pond's surface. Therefore, make-up water (11) is needed.

The cycle itself is closed with respect to the oxygen and carbon and nitrogen is not a component of the cycle. The only need is exergy from the Sun; and according to the Second Law, the entropy flow from the pond exceeds the entropy coming from the falling sun beams.

NUMERICAL RESULTS FOR THE SOFT CYCLE

The best place for this type of plant is in a densely populated belt between 20 and 40 degrees latitude (e.g., Greece, Italy, Spain, Florida, Texas, China, Uzbekistan, Turkey, Israel, etc.). Being clean, such plants might be situated close to consumers. The average solar energy current here equals 250 W/m².

In the pond, the following photosynthesis takes place:

$$x \ CO_2 + x \ 2H_2O + x \ hv \ \rightarrow \ C_xH_{4x} + x \ 2O_2$$

where hv equals the photon energy from the sun. In the combustion chamber, the reaction is just the opposite, i.e.,

$$CH_4 + 2O_2 \rightarrow CO_2 + 2H_2O + 55 \ MJ/kg \ CH_4$$
$$(16) \quad (64) \qquad (44) \qquad (36)$$

One sees here the conversion of solar energy into high concentrated chemical energy in the artificial fuel. This process is a reproduction of the natural fossil fuel formation

FIGURE 2.24 (Continued).

which has been going on for millenniums and which man has the ability to exhaust within a mere century.

The crucial question of the whole project is the efficiency of the photosynthesis. This problem has been considered in detail in reference [16], especially in the comprehensive review of 332 references by K. Zamaraev and V. Parmon which appears on pages 43 through 82 of Ref. [16]. The limiting figure given for this efficiency is 0.15. It is irrelevant to the Carnot factor and depends upon the quantum character of light absorption. All the quanta with an energy less than the threshold one are useless and even those absorbed transfer only a part of their energy. The most reliable data for real calculations can be found in the experimental work of some of the research groups at the University of Moscow [17]. In order to avoid mistakes, some of the pertinent text is translated literally below, i.e.,

> *Since 1980, the development of the solar energy converter referred to as the "biosolar" has been done by the physics faculty, chemistry faculty and some other faculties. Based on this work, the microalgae are planted over a large surface with subsequent methane production. The mineralized elements from the methane tanks are reused by the algae. Carbon is used in the form of CO_2 and delivered in the final product as CH_4. After burning, the CO_2 is restored. Therefore, the global change of CO_2 in the atmosphere is absent. The combustion of fossil fuel, on the other hand, contributes to the greenhouse effect due to the increase of CO_2 in the atmosphere.*

> *The surfaces which are to be covered by photosynthesis plants depend on energy intensity. An efficiency of up to 20% of active radiation is achieved in the laboratory. Certainly, this high figure is not likely under natural conditions. In fact, experience with the massive production of microalgae shows that the real efficiency is between 8 and 10 %.*

> *If the average radiation reaches 240 W/m^2, 40 g of dry substance can be produced in a day from 1 m^2, i.e., 1 mtce of methane could be produced from a surface of 70 km^2.*

> *An outline of "biosolar" is presented in Figure 12 for a mixture of the microalgae <u>Chlorella Pyrenoidosa and Ankistrodesmus Braunii</u> which mutually use the metabolism products.*

> *From the pump, the slurry of biomass in water enters the methane tank where a gas 80 % CH_4, 16 % CO_2, 2 % H_2 and less than 2 % of other constituents is produced.*

The translation above of the work of V. Alexejev and his coauthors is a firm base for proceeding further, although more recent data may be available since photosynthesis investigations are in progress in many countries. However, the author believes that the data in Refs. [16,17] is valid. Thus, the total efficiency of the solar-fuel process is assumed to be 0.10. In the case that new data is found, the results presented here can easily be updated.

The cycle efficiency is the relation of the sum of the three turbines' power less the oxygen consumption divided by the sum of the total enthalpy rises in the combustion chambers, i.e.,

$$\frac{de + fg + hi - D_{O_2}}{cd + ef + gh} = \frac{660 + 640 + 790 - 350}{4480} = 0.39$$

FIGURE 2.24 (Continued).

THE POND'S DIMENSIONS

Land consumption and the depth of the pond are important values which are calculated next. For a solar energy current of 250 W/m², a photosynthesis efficiency of 0.1 and a cycle efficiency of 0.39, the specific power from the pond's surface is 250(.1)(.39) = 10 W/m². Thus, for a large power plant of 1 GW, the land required is 10^9 / 10 = 100 km². In comparison to the Salt Gradient Solar Pond, the SOFT cycle seems to be better because it needs three times less land and the wind is not as detrimental for the SOFT pond as for the salt gradient stability of the latter.

The depth of the pond can not be too great since solar beams could not then penetrate to the bottom of the zone. As is well known, the infrared part of light is absorbed within the upper few centimeters whereas the intensive part of ultraviolet light can reach several meters deep into clear water.

Now, in view of the seasonal variation of the power demand, the SOFT plant is indeed soft since the fuel produced can be stored and the turbine power easily regulated. On a seasonal basis, some fuel is stored during the sunny summer days and CO_2 after the winter demand. If one wishes to store fuel for a year for a 1 GW plant, 2 mt of fuel are needed. If the content of the fuel in water is 1 %, the water volume must be 0.2 km³ and for a surface area of 100 km², the depth 2 m.

The net energy payback time for the SOFT plant seems to be very short because solar energy is not falling on silicon high energy consuming material but on the surface of water in a pond which is much less energy intensive in its construction.

THE AMSTWEG CYCLE

The most dangerous wastes are those of vehicles and not power plants. Therefore, the important task for OSTWEG type cycles (e.g., the types of cycles which have been presented in this paper) is to find a place for them in the automotive industry. Thus, let us consider the Automotive Steam Turbine Without Exhaust Gases (AMSTWEG) cycle for a liquid gas fuelled car. Such a car contains a tank for a propane-butane gas mixture or even compressed methane. Certainly, in the case of methane, the mass of CO_2 effluent exceeds by up to 2.75 times that of the fuel; but the density of CO_2 is greater and, thus, the volumes of fuel and CO_2 are comparable. Hence, neutral liquid CO_2 separated by a soft membrane might fill the space of the consumed fuel.

A recuperator is not needed here since one large expansion in the turbine is used from 1000 °C and 240 bars to 80 °C and 0.5 bars (see Fig. 2). Table 4 shows the nondimensional breakdown of flow rates for the various constituents prior to this expansion.

Table 4 - Nondimensional flow rates for the AMSTWEG cycle.

nondimensional flow rate	before combustion	after combustion
fuel	0.077	0.0
O_2	0.307	0.0
H_2O	0.616	0.789
CO_2	0.0	0.211
total	1.0	1.0

Note that the extra water in the case of this cycle could be used as make-up for the vehicle or elsewhere.

FIGURE 2.24 (Continued).

The cycle efficiency is given by

$$\frac{(4.56 - 2.64)(.875) - 0.330 - 0.087}{4.56 - 0.33} = 0.30$$

This is similar to the efficiency of modern Otto cycle engines. Of course, such engines have exhausts.

In the expansion work calculation, account was taken of the mass share of CO_2 which was 0.211 and which only contributes 0.086 to the total work of expansion due to the low gas constant. The correction factor in total work is $(0.789 + 0.086) = 0.875$.

CONCLUSIONS

The construction of the cycles presented here requires air splitting devices with capacities of up to 100 kg/s of oxygen for power plants and 10 g/s for car engines. A decrease in power consumption is desirable but not necessary. In fact, in spite of the significant power consumption for air splitting and CO_2 liquefaction, all the cycles presented here have efficiencies similar to those of their existing counterparts. The primary benefit of these cycles is, however, evident - *CLEANLINESS*. The disposal of liquid carbon dioxide and the contaminants dissolved in it can be done without harm to the environment.

It would be good to remember here the conclusion edited by W. Chandler of a joint comprehensive investigation which appears in Ref. [19]. It is as follows:

> *What can the world's leaders do to reduce the risk of climatic change? A global climate convention will be necessary to set enforceable goals and provide specific mechanisms for achieving them. A goal of at least holding emissions constant could be pursued with little economic impact. The priorities for specific action include:*
>
> *Setting efficiency standards at cost-effective levels for appliances and automobiles.*
>
> *Shifting taxes to energy resource use.*
>
> *Shifting energy research funds to develop greater end-use efficiency.*
>
> *Shifting energy research to the long-term development of noncarbon fuels.*
>
> *Sharing technology for efficiency with Eastern Europe, the Soviet Union, and developing countries through joint research developing financial institutions, and promoting joint ventures.*

Of course, all of these words are by all means correct but are as insufficient as a prayer. *And all things whatsoever ye shall ask in prayer, ye shall receive* (Matthew 21:22). In order to protect the atmosphere, the immediate engineering action must be to develop energy supplies without emissions to the atmosphere. The cycles presented here might be a step in the right direction. Further calculations and studies are already in progress.

REFERENCES

1. Yantovskii, E. I., *An Attempt to Generalize Exergy Analysis, Referred to as Exergonomics*, Proceedings of the International Conference on the Analysis of Thermal and Energy Systems, Athens, Greece, June 3-6, 1991.

FIGURE 2.24 (Continued).

2. Bockris, J., *guest comment*, <u>Environment Conservation</u>, vol. 17, no. 3, autumn 1990..

3. <u>Gas Turbine World</u>, no. 16, pp. 38-40, December 1988.

4. Stewart, P. B., Munjal, P. K., <u>Journal of Chemical and Engineering Data</u>, 15(1) pp. 67-71, 1970.

5. Zysin, V. A., *Combined Gas-Steam Installations and Cycles*, <u>Gosenergoisdat</u>, in Russian, Moscow, U.S.S.R., 1962.

6. Kantrovitz, A., Sporn, P., *Magnetohydrodynamics - Future Power Process*, <u>Power Magazine</u>, November 1959..

7. Yantovskii, E. I., *Gasdynamic Generator AC*, auth. sert. of March 31, 1959, no. 128542, <u>Bulletin of Inventions</u>, no. 8, 1960.

8. Yantovskii, E. I., *Working Substances for Magnetogasdynamic Electrical Machines*, auth. sert. of September 1959, no. 128950, <u>Bulletin of Inventions</u>, no. 11, 1960.

9. Yantovskii, E. I., Tolmach, E. M., *Magnetohydrodynamic Generators*, in Russian, <u>M. Nauka</u>, 1972.

10. Yantovskii, E. I., *Disk-shaped MHD Generator of Hall Currents*, auth. sert. of June 4, 1965, no. 187177, <u>Bulletin of Inventions</u>, no. 28, 1966.

11. Giovannini, B., Pain, D., *Scientific and Technical Arguments for the Optimal Use of Energy*, <u>Proceedings of the 2nd Annual CMDC Conference</u>, Zürich, Switzerland, December 10-11, 1990.

12. Garcusha, L. K., Schegolev, G. M., *Entropy Diagrams for Combustion Products*, <u>Naukova Dumka</u>, Kiev, U.S.S.R., 1968.

13. Sokolov, E. Ya., Brodianskii, V. M., *Energy Backround of Heat Transformation and Cooling*, <u>M. Energoisdat</u>, in Russian, 1983.

14. Schilling, H. D., Bonn, B., Krauss, U., *Kohlenvergasung*, <u>Verlag Glükauf GMBH</u>, Essen, 1981.

15. Häfele, W. (editor), <u>Energy in a Finite World</u>, Ballinger, N.Y., U.S.A., 1981.

16. Semenov, N. N. (editor), <u>Solar Energy Conversion</u>, Academy of Sciences of the U.S.S.R., Chernogolovka, in Russian, 1981.

17. Alexejev, V. V., Liamin, M. Ya., Shirokova, E. L., *Biomass of Microalgae use for Solar Energy Conversion*, <u>Techno-economic and Ecology Aspects of Ocean Energy Use</u>, TOI Vladivostok, pp. 53-58, in Russian, 1985.

18. Rubin, B. S.,, *Implications of Future Environmental Regulation of Coal-based Electric Power*, <u>Annual Review of Energy</u>, 14:19, 1989.

19. Chandler, W. U., *Carbon Emissions Control Strategies*, <u>Case Studies in International Cooperation</u>, World Wildlife Fund, Baltimore, Maryland, U.S.A., 1990.

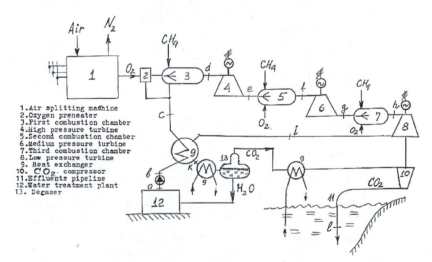

Figure 1 Schema of the GOOSTWEG power plant.

FIGURE 2.24 (Continued).

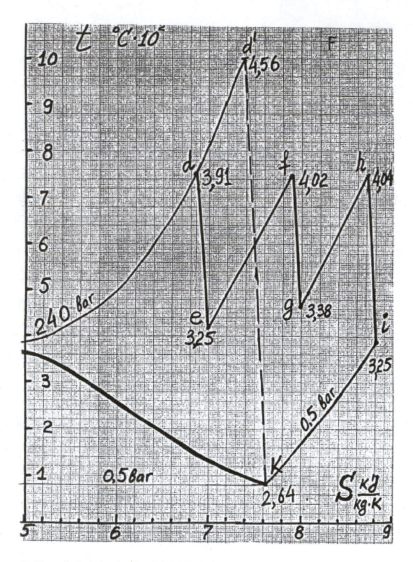

Figure 2 The t-S diagram for the GOOSTWEG cycle.

FIGURE 2.24 (Continued).

A short notice on Zero Emission Power Generation has been published in *Greenhouse Issues*, No. 85 (March 2007, www.ieagreen.org.uk/ghissues.htm). The two workshops are mentioned, the first on January 24, 1997 and next in 1998, organized at Liege University by Professors Yantovsky and Mathieu.

After the second workshop, the event was taken over by IEA GHG and it now forms part of the GHG Technologies conference series.

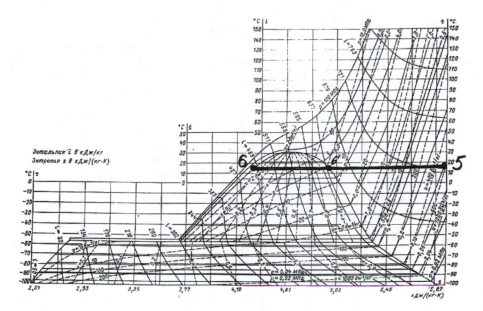

Figure 3 Isothermal liquefaction of CO_2.

FIGURE 2.24 (Continued).

We wish to add here some details of the first workshop from the text of internal Proceedings Workshop: zero emissions power cycles, University of Liege, Jan. 24, 1997.

In attendance were 16 participants — six from the University of Liege, and ten from other places. Here is the list of the ten participants:

AUDUS, H., IEA, Greenhouse R&D Programme, U.K.
DECHAMPS, P., CMI, Belgium
DESIDERI, U., University of Perugia, Italy
KOROBITSYN, M., ECN, the Netherlands
MACOR, A., University of Padua, Italy
McGOVERN, J., University of Dublin, Ireland
MENZ, K., (Mrs), Preusag AG, Germany
MIRANDOLA, A., University of Padua, Italy
PRUSHEK, R., University of Essen, Germany
WOUDSTRA, T., Delft University of Technology, the Netherlands

Here are the concluding remarks by Professor Philippe Mathieu:

The goal of this one-day meeting was to provide all the participants with the most recent works and projects in the field of power generation without any release of pollutants to the atmosphere. Is it not a right of the human being to breath clean air but is it not also his duty to protect his planet?

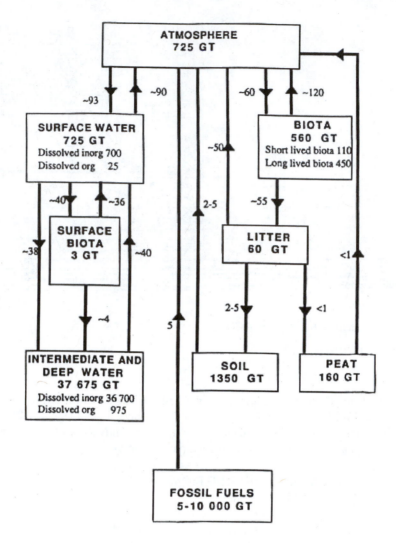

Quantities are in gigatons (1 GT = 10^9 tons) of carbon.

Figure 4 Schematic diagram of the global carbon cycle [15].

FIGURE 2.24 (Continued).

The exchanges of information and ideas were very useful and fruitful and gave the participants the feeling that feasible technical options were available and ready to be implemented at acceptable cost, should mankind (and particularly, the decision makers) consider the greenhouse effect as a critical and urgent issue. Currently, solutions are proposed on paper but there is no political and social will for the undertaking and, consequently, no economic incentive to do so.

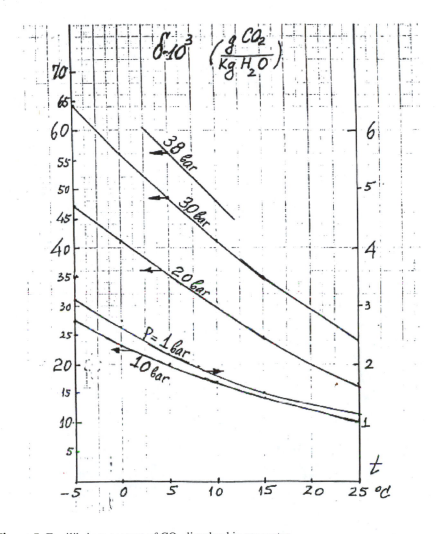

Figure 5 Equilibrium content of CO_2 dissolved in sea water.

FIGURE 2.24 (Continued).

Professor E. Yantovsky gave the background of the workshop, with a reminder that it is now scientifically proven that the greenhouse effect exists. He gave the basic principles of conversion of any power generating system into a zero emissions one, namely the use of a CO_2 cycle, the use of oxygen for the combustion and hence, the need for an air separation unit, the use of CO_2 turbomachinery, the separation of water and excess CO_2, and extraction of that CO_2 in liquid state for transportation and disposal.

The research team from the University of Liege delivered the results of their modelling and optimization of new and original zero emissions power plants.

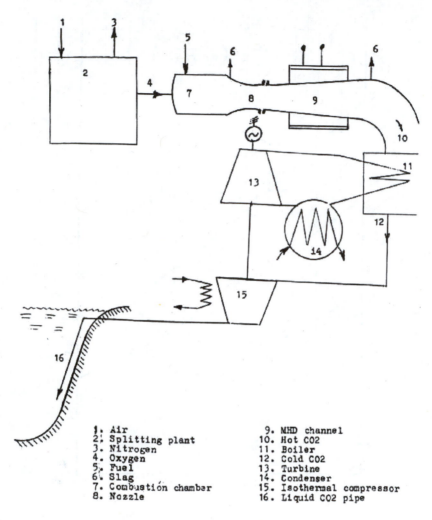

Figure 6 Schema for the COMHDWEG power plant.

FIGURE 2.24 (Continued).

Raoul Nihart showed that a CO_2 power cycle comprised of a Rankine-like cycle on the top of a Brayton CO_2 cycle and operating in compliance with technical and material related constraints can reach efficiencies above 50% with natural gas as a fuel.

Yann Greday provided us with a CO_2 cycle fuelled with biomass produced from microalgae by the solar radiation and reinjected CO_2 in a pond. Such a solar-based system has a payback time much lower than in PV cells and is easy to build using technically proven components.

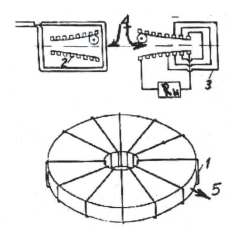

```
1. Winding for magnetic field creation
2. Ring electrodes
3. Electrode connection
4. Plasma from combustion chamber
5. Gas outflow
```

Figure 7 Disk-shaped radial flow MHD generator.

FIGURE 2.24 (Continued).

Rafael Lhomme explained how to integrate a MagnetoHydroDynamic generator nozzle into a fossil fuel fired zero emissions CO_2 cycle. Efficiency from 49 to 54% can be obtained using a well-known technology, including ceramic heat exchangers and a superconductive magnet. This could be an example for use of existing MHD facilities, notably in the former-U.S.S.R.

Concluding this system optimization, Philippe Mathieu talked about the possibility of the use of existing industrial gas turbines for operation on CO_2 instead of air. On the basis of a very simplified one-dimensional model, it appears that a reduction of the rotational speed by 20% gives a very good similarity of CO_2 and air flows through the turbomachine. Could an air-based gas turbines built in the U.S. at 60 Hz be used in Europe with CO_2 at 50 Hz as such or with slight modifications? This has to be tested and more accurate models (with cooling) should be developed to answer that important question.

G. Göttlicher and R. Pruschek from the University of Essen (Germany) presented a very exhaustive comparison of coal- and gas-based power plants with CO_2 removal from the performance, specific residual CO_2 emissions, and cost points of view.

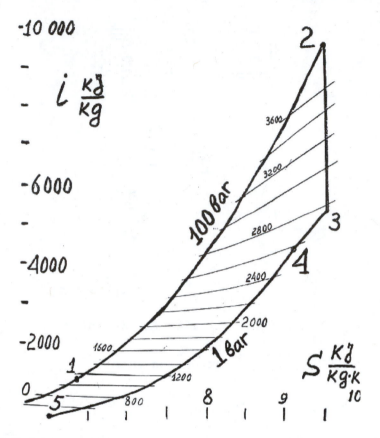

Figure 8 Enthalpy-entropy diagram for the slurry combustion in pure oxygen.

FIGURE 2.24 (Continued).

U. Desideri and R. Corbelly from the University of Perugia (Italy) presented an analysis of performance and cost of CO_2 capture in small-sized power plants. Calculations were focused on a Cheng cycle with cogeneration and chemical CO_2 scrubbing.

To conclude the day, Jim McGovern from the University of Dublin (Ireland) gave his view on the role of computer simulation in the context of zero emissions power plants.

The general conclusions of the workshop are as following:

For every known gas-steam turbine combined cycle (natural gas, integrated gasification, fluidized bed, MHD) exists its zero emissions quasi-combined cycle counterpart (ZEQC), CO_2 being used as the working substance.

The new elements of ZEQC are: air separation unit for oxygen production, CO_2 high- and low-pressure turbines, inherent production of liquid CO_2 at 60 bar and its injection underground or in the ocean.

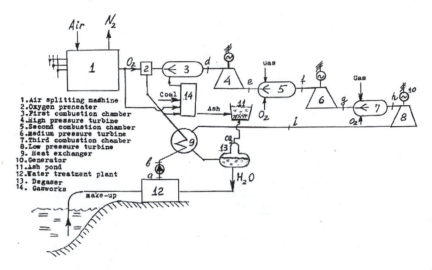

Figure 9 Outline of the coal-oxygen steam turbine power plant without exhaust gases and ash neutralization by carbonic acid.

FIGURE 2.24 (Continued)

The efficiency penalty due to oxygen production power is offset by smaller compression work of CO_2 than with air. Therefore, ZEQS efficiency is the same or higher than its originator.

If an ordinary gas turbine, designed for 3600 rpm, were running on CO_2 at 3000 rpm, it would have similar hydrodynamics and isentropic efficiency for a given power output.

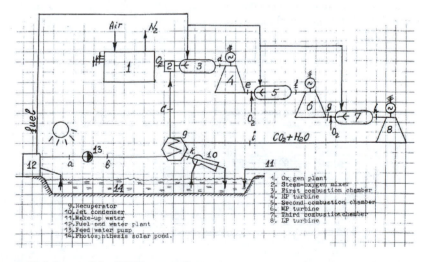

Figure 10 Schema of the SOFT power plant.

FIGURE 2.24 (Continued).

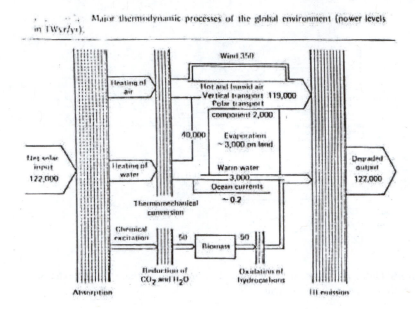

Figure 11 Global energy flows.

FIGURE 2.24 (Continued).

There exist a number of economically attractive ZEQC applications, including
combination of power generation with incineration or oil recovery, especially in Europe.

The Department of Nuclear Engineering and Power Plants of Liege University
is ready to initiate and take part in the design and construction of a demonstration ZEQC power plant.

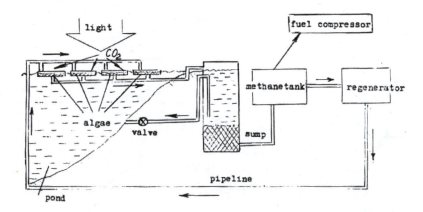

Figure 12 Diagram of the "biosolar" scheme.

FIGURE 2.24 (Continued).

The leadership in the zero emissions power plant construction will create many jobs in Europe.

During lunch, Philippe Mathieu declared that the participants at the workshop constituted the first permanent workgroup on zero emissions power generation and promised to organize the event annually.

REFERENCES

Akinfiev, N., McGovern, J., and E. Yantovsky. 2005. Zero Emissions Power Generation with CO_2 Reduction by Fayalite, *Int. J. Thermod.*, 8 (3): 155–157.

Anderson, R.E., Doyle, S.E., and K.L. Pronske. 2004. Demonstration and Commercialization of Zero-Emission Power Plants, presented at *29th International Technical Conference on Coal Utilization & Fuel Systems*, April 18–22, Clearwater, FL, U.S.A.

Anderson, R.E. et al. 2008. Adapting gas turbines to zero-emissions oxy-fuel power plants, in *Proc. ASME Turbo Expo 2008 GT 2008*, June 9–13, Berlin.

Anna, D. DOE Advances Oxycombustion for Carbon Management, *DOE National Energy Technology Laboratory*, www.fe.doe.gov/news/techlines/2005/tl_oxycombustion_award.html, (accessed Nov. 17, 2005).

Ausubel, J. 2004. Big green energy machines, *The Industrial Physicist*, Oct/Nov, 21.

Beichel, R. 1996. Reduced pollution power generation system. U.S. Patent 5,715,673, Feb.

Berry, G.F. and A.M. Wolsky. 1986. Modelling Heat Transfer in an Experimental Coal-Fired Furnace When CO_2/O_2 Mixtures Replace Air, Paper 86-WA/HT-51, ASME WAM, Anaheim, CA, U.S.A.

Bolland, O. 2004a. Power cycles with CO_2 capture, *NTNU*, Tep. 9, www.ept.ntnu.no/fag/tep9/inhold/CO2_tep9.pdf, (accessed Oct. 15, 2006).

Bolland, O. 2004b. CO_2 capture technologies—an overview, in *Proc. of The Second Trondheim Conference on CO_2 Capture, Transport and Storage*, Oct. 25–26, 2004, Trondheim, Norway.

Bolland, O. and S. Saether. 1992. New concepts for natural gas fired power plants which simplify the recovery of carbon dioxide, *En. Conv. Mgmt.*, 33 (5–8): 467–475.

Bouwmeester, H. and A. Burggraaf. 1996. Dense Ceramic Membranes for Oxygen Separation, *CRC Handbook of Solid State Electrochemistry*, CRC: Boca Raton, Chap. 11.

Bredesen, R., Jordal, K., and O. Bolland. 2004. *High temperature membranes in power generation with CO_2 capture*, www.zero.no/fossil/CO2/200409121620/200409121627, (accessed Jan. 15, 2006).

Degtiarev, V.L. and V.P. Gribovsky. 1967. "Carbon dioxide semi-closed power plant," published in Russ. Bull. Inventions No. 8, F01k13/00, Co1b 3/00, Nov. 12, 1971, Author certif. July 28, USSR No. 295 897.

De Ruyck, J. 1992. Efficient CO_2 capture through a combined steam and CO_2 gas turbine cycle, *En. Conv. Mgmt.*, 33 (5–8): 397–404.

Dyer, P., Richards, R., Russek, S., and D. Taylor. 2000. Ion transport membrane technology for oxygen separation and syngas production, *Solid State Ionics*, (134): 21–33.

Foy, K. and J. McGovern. 2005. Comparison of Ion Transport Membranes, in *Proc. 4th Annual Conference on Carbon Capture and Sequestration*, May 2–5, Alexandria, VA, U.S.A., Paper 111.

Göttlicher, G. 1999. Energetik der Kohlendioxidrückhaltung, in *VDI-Fortschrittberichte, Reihe 6*, No. 421, Düsseldorf, (in German).

Göttlicher, G. 2003. *Process comparison* 2003, http://eny.hut.fi/education/courses/Ene-47_200-2003/GHG2003_course_material+3-G%F6ttlicher-CO_2-Capture.pdf, (accessed Aug. 11, 2006).

Gupta, M., Coyle, I., and K. Thambimuthu. 2003. CO_2 capture technologies and opportunities in Canada, in *Proc. 1st Canadian CC&S Technology Workshop*, Sept. 18–19, Calgary.

Heitmeir, F. et al., 2003.The Graz cycle — A zero emission power plant of highest efficiency, in *XXXV Kraftwerkstechnisches Kolloquium*, Sept 23–24, Dresden, Germany, http://www.graz-cycle.tugraz.at/ (accessed Sept. 14, 2006).

Hochstein, D.P. 1940. Carbon dioxide power cycle, *Soviet Boiler and Turbine Construction*, No.10: 420–423, (in Russian).

Hoffman, A. and E. Feher. 1971. 150 kWe Supercritical Closed Cycle System, *ASME Trans.*, Ser. *A*, 73 (1): 104–109.

Holt, T. and E. Lindeberg. 1992. Thermal power—without greenhouse gases and with improved oil recovery, *En. Conv. Mgmt.*, 33 (5–8): 595–602.

International Energy Agency Working Party on Fossil Fuels, Technology Status Report: Zero Emission Technologies for Fossil Fuels., 2002, http://www.cslforum.org/documents/TSRMay2002.pdf, (accessed Dec. 9, 2005).

Ishida, M. and H. Jin. 1998. "Greenhouse gas control by a novel combustion: no separation equipment and energy penalty," in *4th Int. Cong. GHGT*, Interlaken, ENT-13.

Jericha, H. 1985. Efficient Steam Cycles with Internal Combustion of Hydrogen and Stoichiometric Oxygen for Turbines and Piston Engines, in *CIMAC Conference Papers*, Oslo, Norway.

Jericha, H. et al., 1995. CO_2 Retention Capability of CH_4/O_2 — Fired Graz Cycle, in *Proc. CIMAC*, Interlaken, Switzerland.

Jericha, H. Lukasser, A., and W. Gatterbauer. 2000. Der "Graz Cycle" für Industriekraftwerke gefeuert mit Brenngasen aus Kohle- und Schwerölvergasung (in German), *VDI Berichte* 1566, VDI Conference Essen, Germany.

Jericha, H. et al. 2004. Design Optimization of the Graz Cycle Prototype Plant. *ASME Trans.*, Ser. *A*, 126 (3): 733–740.

Keller, C. and R. Strub. 1968. "The gas turbine for nuclear power reactors," presented in *VII World Energy Congress*, Moscow, Paper 167.

Knoche, K.F. and H. Richter. 1968. Verbesserung der Reversibiltaet von Verbrennungs-procesen, *BWK*, (5): 205–210, (in German).

Leithner, R. 2005. Energy conversion processes with intrinsic CO_2 separation, *Trans. of the Society for Mining, Metallurgy, and Exploration*, 318: 161–165.

Levin, L., Nesterovski, I., and E. Yantovsky. 2003. "Zero emission Rankine cycle with separate ion transport membrane combustor," presented at *7th Int. Conf. for a Clean Environment*, July 7–10, Lisbon, Portugal.

Lorentzen, G. and J. Pettersen. 1990. Power process development for northern climate, in *Eurogas'90. Proc. of Conf. on Natural Gas*, May 28–30, Trondheim, 451–462.

Marchetti, C. 1979. "Constructive solutions to the CO_2 problem," in *Man's Impact on Climate*, Elsevier: New York.

Marin, O., Bourhis, Y., Perrin, N., Di Zanno, P., Viteri, F., and R. Anderson. 2005. *High Efficiency, Zero Emission Power Generation Based on a High-Temperature Steam Cycle*, www.cleanenergysystems.com/2005/mediakit/AL_CESClearwaterPaper.pdf, (accessed Dec. 9, 2005).

Mathieu, P. 1994. "The use of CO_2 Gas Turbines," presented at *Power-Gen. Europe '94 Conf.*, Cologne, May 17–19, Germany.

Mathieu, P. 1998. Presentation of an innovative Zero-Emission cycle for mitigation the global climate change, *Int. Jour. of Applied Thermodyn.*, vol. 1, No.1–4.

Mathieu, P., Dubuisson, R., Houyou, S., and R. Nihart. 1999. "Combination of near zero emission power cycles and CO_2 sequestration," presented at *5th Int. Conf. on Technologies and Combustion for a Clean Environment,* July, Lisbon.

Mathieu P. and F. Desmaret. 2001. "Integration of a high temperature fuel cell (SOFC) in a near zero CO_2 emission power cycle," *ECOS 2001,* May, Istanbul, Turkey.

Mathieu, P. and F. van Loo. 2005. Modeling of an IGCC Plant based on oxy-fuel combustion combined cycle, in *Proc. ECOS 2005*, June 20–22, Trondheim, Norway.

Nakayama, S. et al., 1992. Pulverized coal combustion in O_2/CO_2 mixture on a power plant for CO_2 recovery, *Energy Conv. Mgmt.*, 33(5–8): 379–386.

Pak, P.S., Nakamura, K., and Y. Suzuki. 1989. "Closed dual fluid gas turbine power plant without emission of CO_2 into the atmosphere," in *IFAC/IFORS/IAEE Int. Symp. on Energy Systems Management and Economics*, Oct., 249–254., U.S. Pat. 5,247,791, Sept. 28, 1993.

Pechtl, P. 1991. "CO_2 emissionsminderung," *Erdol und Kohle Petrochemie*, H. 4 (Apr.): 159–162, (in German).

Pham, A.Q. 1997. Making Liquid Fuels from Natural Gas, www.vacets.org/vtic97/qapham. html, (accessed Dec. 10, 2006).

Renz, U. 2004. "Entwicklung eines CO_2-emissionsfreien Kohleverbrennungs-processes zur Stromerzeugung in einem Verbundvorhaben der RWTH Aachen," in *XXXVI. Kraftwerkstechnisches Kolloquium: Entwicklungpotentiale fuer Kraftwerke mit fossile Brennstoffen*, Oct. 19–20, Dresden, (in German).

Renz, U., Kneer, R., Abel, D., Niehuis, R., Maier, H., Modigell, M., and N. Peters. 2005. "Entwicklung eines CO_2-emissionsfreien Kohleverbrennungs-processes zur Stromerzeugung," in *CCS-Tagung Juelich,* Nov. 10–11, RWTH Aachen, (in German).

Romano, M. et al., 2005. Decarbonized Electricity Production from Coal by means of Oxygen Transport Membranes, in *4th Annual Conference on Carbon Sequestration*, May 2–5, Alexandria, VA, 1402.

Ruether, J., Le, P., and Ch. White. 2000. A zero-CO_2 emission power cycle using coal. *Technology,* 7S: 95–101.

Sanz, W. et al., 2004. Thermodynamic and Economic Investigation of an Improved Graz Cycle Power Plant for CO_2 Capture, ASME Paper GT2004-53722, ASME Turbo Expo, Vienna, Austria.

Sanz, W., Jericha, H., Luckel, F., Göttlich, E., and F. Heitmeir. 2005. A Further Step Towards a Graz Cycle Power Plant for CO_2 Capture, ASME Paper GT2005-68456, ASME Turbo Expo, Reno–Tahoe, Nevada, U.S.A.

Selimovic, F. 2005. "Modelling of Transport Phenomena in Monolithic Structures related to CO_2-free power process," Lund University of Technology, Sweden, http://130.235. 81.176/~ht/documents/lic-pres.pdf (accessed Nov. 9, 2006).

Shockling, L., Huang, K., and G. Christie. 2001. "Zero Emission Power Plants Using Solid Oxide Fuel Cell and Oxygen Transport Membranes," *Vision 21 Program Review meeting*, Nov. 6, U.S. DOE, NETL, http://www.osti.gov/bridge/servlets/purl/ 832845-EDQk0q/native/832845.pdf, (accessed Dec. 12, 2005).

Steinberg, M. 1981. A carbon dioxide power plant for total emission control and enhanced oil recovery, Brookhaven National Laboratory, Upton, NY 11973, *BNL,* 30046, August.

Steinberg, M. 1992. History of CO_2 greenhouse gas mitigation technologies, *Energy Conv. Mgmt.*, 33 (5–8): 311–315.

Sundkvist, S., Griffin T., N. Thorshaug. 2001. "AZEP-Development of an Integrated Air Separation Membrane-Gas Turbine," *Second Nordic Minisymposium on CO_2 Capture*, Oct. 26, Chalmers University, Göteborg.

Sundkvist, S. et al., 2004. AZEP gas turbine combined cycle power plants, in *GHGT-7*, University Regina, Sept. 5–9, Vancouver, Canada,//uregina.ca/ghgt7/PGF/papers/ peer/079.pdf, (accessed Dec. 7, 2005).

Sundkvist, S., Eklund, H., and T. Griffin. 2004. AZEP — an EC funded Project for Development of a CCGT Power Plant without CO_2 Emissions, in CAME-GT *2nd Int. Conference on Gas Turbine Technologies*, April 29–30, Bled, Slovenia, www.came-gt.com/. 2InternatConf/Tech%20Session%203/SGSundkvist.pdf, (accessed Dec. 7, 2005).

Tomski, P. 2003. Zero Emission Technologies for Fossil Fuels Technology Status Report. App. I: R&D Visions (Request for Project Updates), Preliminary Draft Update: June, www. cslforum.org/documents/TSRAppendix2003.pdf (accessed Dec. 9, 2005).

van der Haar, M. 2001. Mixed-conducting perovskite membranes for oxygen separation, Ph.D. thesis, University of Twente, Enschede.

van Steenderen, P. 1992. "Carbon dioxide recovery from coal and natural gas fired combined cycle power plants by combustion in pure oxygen and recycled carbon dioxide," *COMPRIMO BV Consulting Services,* Amsterdam, the Netherlands.

VGB PowerTech, 2005. "CO_2 Capture and Storage," VGB PowerTech Service GmbH., Essen, August 25, www.vgb.org/data.o/vgborg_/Fachgremien/Umweltschutz/ VGB%20 Capture%20and%20Storage.pdf, (accessed Dec. 12, 2006).

Wall, G., Yantovsky, E., Lindquist, L., and J. Tryggstad. 1995. A Zero Emission combustion power plant for enhanced oil recovery, *Energy—The Int. Jour.,* vol. 20 (8): 823–828.

Wolsky, A. 1985. A new approach to CO_2 recovery from combustion, in *Proc. of a Workshop: Recovering Carbon Dioxide from Man-Made Sources,* Feb. 11–13, Pacific Grove, CA, Argonn Nat. Lab. Report ANL/CNSV-TM-166, 76–81.

Yantovsky, E. 1991. "The thermodynamics of fuel-fired power plants without exhaust gases," in *Proc. of World Clean Energy Conf. CMDC,* Nov. 4–7, Geneva, 571–595.

Yantovsky, E., Zvagolski, K., and V. Gavrilenko. 1992. Computer exergonomics of power plants without exhaust gases, *En. Conv. Mgmt.,* 33 (5–8): 405–412.

Yantovsky, E. and V.L. Degtiarev. 1993. Internal combustion carbon dioxide power cycles without exhaust gases, in *Proc. of Int. Conf. ENSEC'93,* July 5–9, Cracow, Poland, 595–602.

Yantovsky, E., Wall, G., Lindquist, L., Tryggstad, J., and R. Maksutov. 1993. Oil Enhancement Carbon Dioxide Oxygen Power Universal Supply (OCDOPUS project), *En. Conv. Mgmt.,* 34 (9–11): 1219–1227.

Yantovsky, E. et al., 1994a, Exergonomics of an EOR (Ocdopus) project, *Energy — The Int. Jour.,* No. 12, 1275–1278.

Yantovsky, E., Zvagolsky, K., and V. Gavrilenko, 1994b, The COOPERATE power cycle, in *Proc. WAM, ASME, AES,* vol. 33, 105–112.

Yantovsky, E. 1994c, *Energy and Exergy Currents,* NOVA Sci.: NY.

Yantovsky, E. 1996. Stack downward. Zero Emission Fuel-fired power plant concept, *Energy Conversion Management,* 37 (6–8): 867–877.

Yantovsky, E. and V. Kushnirov, 2000b, "The Convergence of Oil and Power on Zero Emission Basis," *OILGAS European Magazine,* No. 2.

Yantovsky, E. 2000c. "Gas Power Zero Emission Convergency Complex," *OILGAS European Magazine,* No. 4.

Yantovsky, E., Gorski, J., Smyth, B., and J.E. ten Elshof. 2002. ZEITMOP Cycle (Zero Emission Ion Transport Membrane Oxygen Power), *Proc. Int. Conf. ECOS 2002,* July 3–5, Berlin, 1153–1160.

Yantovsky, E., Gorski, J., Smyth, B., and J. ten Elshof. 2003a. Zero Emission Fuel-Fired Power Plant with Ion Transport Membrane, in *2nd Annual Conf. On Carbon Sequestration,* May 5–8, Alexandria, VA, U.S.A.

Yantovsky, E. and M. Shokotov. 2003b. ZEMPES (Zero Emission Piston Engine System), in *2nd Annual Conf. On Carbon Dioxide Sequestration*, May 5–8, Alexandria, VA, U.S.A.

Yantovsky, E., Shokotov, M., McGovern, J., and V. Vaddella. 2004. Zero emission membrane piston engine system (ZEMPES) for a bus, in *Proc. VAFSEP*, July 6–9, Dublin, 129–133, www.netl.doe.gov/publications/proceedings/01/vision21/v211-5.pdf.

Yantovsky, E., Shokotov, M., McGovern, J., Shokotov, V., and K. Foy. 2005. Elaboration of Zero Emission Membrane Piston Engine System (ZEMPES) for propane fuelling, in *4th Carbon Sequestration Conf.*, May 2–5, Alexandria, VA, U.S.A, Paper 109.

3 Zero Emissions Quasicombined Cycle with External Oxygen Supply

3.1 CARBON DIOXIDE—THERMODYNAMIC PROPERTIES, PURE AND MIXTURES

From an engineering point of view, carbon dioxide has been of great importance in many industrial processes and applications for several decades. It is used mainly as a solvent for supercritical fluid extraction in the chemical industry and for enhanced oil well recovery. Some current developments have shown its suitability for new transcritical cycles and ecologically safe air conditioning and heat pump systems. The application of carbon dioxide as a refrigerant (R-744) in automotive air conditioning systems has grown since the pioneering works of G. Lorenzen at the end of the last century.

As the main component of effluent gases from the combustion of coal or hydrocarbon-based fuels, CO_2 is responsible for atmospheric emissions and the GHG effect. Its capture and utilization play a main role in the future of energy conversion systems. Another method proposed for capturing CO_2 in power generation is through the semi-closed O_2/CO_2 gas turbine cycle, which burns an oxy-fossil fuel mixture utilizing recycled CO_2 as a diluting gas.

The thermodynamic and transport properties of fluid-state carbon dioxide are widely recognized in many open sources (http://airliquide, 2008; www.chemicalogic, 2008; Span and Wagner, 1996; Vesovic et al., 1990). Several important parameters explaining its material compatibility, major hazards, and safety data sheets are given in thermodynamics handbooks (Gorski, 1997).

Most thermodynamic properties (Span and Wagner, 1996), as well as the solubility of carbon dioxide in aqueous solutions (Kerrick and Jacobs, 1981; Mäder, 1991), can be described by using the appropriate PVT-relations or equations of state (EOS) (Span, 2000). Phase relations can be derived from phase diagrams for H_2O and CO_2, as well as the CO_2-H_2O system, shown in Figure 3.1.

An extensive review of available data on CO_2 thermodynamic properties is presented in the well-known *Carbon Dioxide Standard* (Span and Wagner, 1996). Equation of state developed by Span and Wagner (1996) covers the fluid region from the triple point temperature up to 826.85°C and pressures up to 8000 bars. Span and Wagner (1996) developed a new EOS for the representation of the thermodynamic properties of carbon dioxide, which is an empirical representation of the fundamental

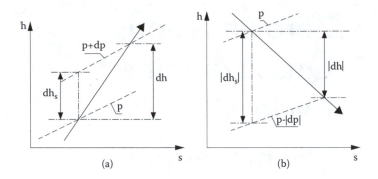

FIGURE 3.1 Elementary polytropic compression (a) and expansion (b). (From Gorski, 1997. With permission.)

equation explicit in the Helmholtz free energy A. The stated fundamental equation is expressed with the two independent variables: density ρ and temperature T in the nondimensional form. According to recent thermodynamic methods, the dimensionless Helmholtz energy $\phi = A/RT$ is usually spliced into an ideal gas part ϕ° and residual, real gas part ϕ^r:

$$\phi \overset{def}{=} \frac{A(\delta, \tau)}{RT} = \phi^0(\delta, \tau) + \phi^r(\delta, \tau) \tag{3.1}$$

where $\delta = \rho/\rho_c$ is the reduced density, i.e., the density normalized to the critical density ρ_c, and $\tau = T_c/T$ is the inversed reduced temperature, i.e., the inverse of temperature normalized to critical temperature T_c (the subscript c denotes the critical values of fluid density and temperature).

Here, the Helmholtz energy of one component fluid is described as a function of density and temperature. However, this is just one form of a fundamental equation and, in fact, all thermodynamic properties of pure carbon dioxide can be calculated by combining derivatives of this equation (see Table 3 in Span and Wagner, 1996 and Table 3.1).

A simpler approach for calculation of fluid properties can be used based on the virial compressibility derivatives (VCD) symbolism (Gorski, 1997).

Some principal relations comparing the use of VCD and classical notation (Span and Wagner, 1996) for calculation of selected thermodynamic parameters are given in Table 3.1. The use of VCD notation allows simplifying the mathematical expressions for calculating all important derivatives of the thermal parameters. For example:

$$\left(\frac{\partial p_r}{\partial T_r} \right)_\rho \equiv -\tau^2 \left(\frac{\partial p_r}{\partial \tau} \right)_\delta = \frac{z_{\tau\delta}\delta}{z_c} \equiv \frac{\delta}{z_c}(\phi_\delta - \tau\phi_{\delta\tau}),$$

$$\left(\frac{\partial^2 p_r}{\partial T_r^2} \right)_T \equiv -\frac{\delta\tau^2}{z_c} \left(\frac{\partial z_{\tau\delta}}{\partial \tau} \right)_\delta \equiv \frac{\delta^2\tau^3}{z_c}\phi_{\delta\delta\tau},$$

TABLE 3.1
Thermodynamic Functions Represented in Terms of Reduced Helmholtz and VCDs

Parameter and Definition	Reduced Helmholtz Function	Virial Compressibility Derivatives
Pressure $p(T,\tilde{V}) = -\left(\dfrac{\partial \tilde{F}}{\partial \tilde{V}}\right)_T$	$\dfrac{p\tilde{V}}{RT} = 1 + \delta\phi_\delta^r$	$p = \dfrac{z\tilde{R}T}{\tilde{V}} = zpRT$
Enthalpy $\tilde{H}(T,p) = \tilde{F} - T\left(\dfrac{\partial \tilde{F}}{\partial T}\right)_v - \tilde{V}\left(\dfrac{\partial \tilde{F}}{\partial \tilde{V}}\right)_T$	$\dfrac{\tilde{H}(\delta,\tau)}{RT} = 1 + \tau\left(\phi_\tau^{id} + \phi_\tau^r\right) + \delta\phi_\delta^r$	$\dfrac{\tilde{H}}{RT} = \dfrac{\tilde{H}^{id}}{RT} + T\int_0^\rho\left[\left(\dfrac{\partial z}{\partial T}\right)_\rho\right]\dfrac{d\rho}{\rho} + (1-z)$
Entropy $\tilde{S}(T,\tilde{V}) = -\left(\dfrac{\partial \tilde{F}}{\partial T}\right)_{\tilde{V}}$	$\dfrac{\tilde{S}(\delta,\tau)}{R} = \tau\left(\phi_\tau^{id} + \phi_\tau^r\right) - \phi^{id} + \phi^r$	$\dfrac{\tilde{S}}{R} = \dfrac{\tilde{S}^{id}}{R} + \int_0^\rho\left(z_{Tv} - 1\right)\dfrac{d\rho}{\rho} - \ln z$
Internal energy $\tilde{U}(T,\tilde{V}) = \tilde{F} - T\left(\dfrac{\partial \tilde{F}}{\partial T}\right)_{\tilde{V}}$	$\dfrac{\tilde{U}(\delta,\tau)}{RT} = \tau\left(\phi_\tau^{id} + \phi_\tau^r\right)$	$\dfrac{\tilde{U}}{RT} = \dfrac{\tilde{U}^{id}}{RT} + T\int_0^\rho\left[\left(\dfrac{\partial z}{\partial T}\right)_\rho\right]\dfrac{d\rho}{\rho}$
Specific heat (at $V = const.$) $\tilde{C}_v(T,\tilde{V}) = \left(\dfrac{\partial \tilde{U}}{\partial T}\right)_{\tilde{V}}$	$\dfrac{\tilde{C}_v(\delta,\tau)}{R} = -\tau^2\left(\phi_{\tau\tau}^{id} + \phi_{\tau\tau}^r\right)$	$\dfrac{\tilde{C}_v - \tilde{C}_v^{id}}{R} = \dfrac{\tilde{C}_v^{id}}{R} + T\int_0^\rho\left[\left(\dfrac{\partial z_{Tv}}{\partial T}\right)_\rho\right]\dfrac{d\rho}{\rho}$
Specific heat (at $p = const.$) $\tilde{C}_p(T,p) = \left(\dfrac{\partial \tilde{H}}{\partial T}\right)_p = \tilde{C}_v(T,\tilde{V}) + \left(\dfrac{\partial p}{\partial T}\right)_{\tilde{V}}^2 \Big/ \left(\dfrac{\partial p}{\partial \tilde{V}}\right)_T$	$\dfrac{\tilde{C}_p(\delta,\tau)}{R} = -\tau^2\left(\phi_{\tau\tau}^{id} + \phi_{\tau\tau}^r\right) + \dfrac{\left(1 + \delta\phi_\delta^r - \delta\tau\phi_{\delta\tau}^r\right)^2}{1 + 2\delta\phi_\delta^r + \delta^2\phi_{\delta\delta}^r}$	$\dfrac{\tilde{C}_p}{R} = \dfrac{\tilde{C}_v}{R} + \dfrac{z_{Tv}^2}{z_{vT}}$

(Continued)

TABLE 3.1 (CONTINUED)
Thermodynamic Functions Represented in Terms of Reduced Helmholtz and VCDs

Parameter and Definition	Reduced Helmholtz Function	Virial Compressibility Derivatives
Sound velocity $a_s(T,\tilde{V}) = -\tilde{V}\left(\dfrac{\partial p}{\partial \tilde{V}}\right)_s^{1/2}$	$\dfrac{a_s^2(\delta,\tau)}{RT} = 1 + 2\delta\phi_\delta^r + \delta^2\phi_{\delta\delta}^r - \dfrac{\left(1+\delta\phi_\delta^r - \delta\tau\phi_{\delta\tau}^r\right)^2}{\tau^2\left(\phi_{\tau\tau}^{id} + \phi_{\tau\tau}^r\right)}$	$\dfrac{a_s^2}{RT} = \gamma z_{vT}$
Joule-Thompson coefficient $\mu_{JT}(T,p) = \mu_{JT}(T,\tilde{V}) = \left(\dfrac{\partial T}{\partial p}\right)_h$	$\mu_{JT}\cdot\dfrac{\tilde{R}}{\tilde{V}} = \dfrac{-\left(\delta\phi_\delta^r + \delta^2\phi_{\delta\delta}^r + \delta\tau\phi_{\delta\tau}^r\right)}{\left(1+\delta\phi_\delta^r - \delta\tau\phi_{\delta\tau}^r\right)^2 - \tau^2\left(\phi_{\tau\tau}^{id}+\phi_{\tau\tau}^r\right)\left(1+2\delta\phi_\delta^r+\delta^2\phi_{\delta\delta}^r\right)}$	$\mu_{JT}Rp = \dfrac{\tilde{R}}{\tilde{C}_p}\left(\dfrac{z_{Tv}}{z_{vT}} - 1\right)$
Compressibility factor $z(T,\tilde{V}) = \dfrac{p\tilde{V}}{RT}$	$z(\delta,\tau) = 1 + \delta\phi_\delta^r$	$z(T,\rho)=z(T,\tilde{V}) = \dfrac{p}{\rho RT} = \dfrac{p\tilde{V}}{RT}$

Abbreviations: $\phi_\delta = \left(\dfrac{\partial\phi}{\partial\delta}\right)_\tau$, $\phi_{\delta\delta}=\left(\dfrac{\partial^2\phi}{\partial\delta^2}\right)_\tau$, $\phi_\tau=\left(\dfrac{\partial\phi}{\partial\tau}\right)_\delta$, $\phi_{\tau\tau}=\left(\dfrac{\partial^2\phi}{\partial\tau^2}\right)_\delta$, $\phi_{\delta\tau}=\left(\dfrac{\partial^2\phi}{\partial\delta\partial\tau}\right)$

Source: Gorski, J., 1997.

$$\left(\frac{\partial p_r}{\partial \rho_r}\right)_T \equiv \left(\frac{\partial p_r}{\partial \delta}\right)_\tau = \frac{z_{\delta\tau}}{z_c\tau} \equiv \frac{\delta}{z_c\tau}(2\phi_\delta + \delta\phi_{\delta\delta}),$$

$$\left(\frac{\partial^2 p_r}{\partial \rho_r^2}\right)_T \equiv \left(\frac{\partial^2 p_r}{\partial \delta^2}\right)_\tau = \frac{1}{z_c\tau}\left(\frac{\partial z_{\delta\tau}}{\partial \delta}\right)_\tau \equiv \frac{\delta}{z_c\tau}(2\phi_\delta + 4\delta\phi_{\delta\delta} + \delta\phi_{\delta\delta\delta}),$$

$$\left(\frac{\partial^2 p_r}{\partial T_r \partial \rho_r}\right) \equiv \frac{1}{z_c}\left[\frac{\partial(\delta z_{\tau\delta})}{\partial \delta}\right]_\tau = \frac{\delta}{z_c\tau}(2\phi_\delta + \delta\phi_{\delta\delta} - 2\tau\phi_{\delta\tau} - \delta\tau\phi_{\delta\delta\tau}).$$

where $z_c = \frac{p_c}{\rho_c R T_c}$, is the compressibility factor at thermodynamic critical point and

$$\phi_\delta = \phi_\delta^0 + \phi_\delta^r = \left(\frac{\partial\phi}{\partial\delta}\right)_\tau, \quad \phi_\delta^0 = \left(\frac{\partial\phi^0}{\partial\delta}\right)_\tau, \quad \phi_\delta^r = \left(\frac{\partial\phi^r}{\partial\delta}\right)_\tau,$$

$$\phi_\tau = \phi_\tau^0 + \phi_\tau^r = \left(\frac{\partial\phi}{\partial\tau}\right)_\delta, \quad \phi_\tau^0 = \left(\frac{\partial\phi^0}{\partial\tau}\right)_\delta, \quad \phi_\tau^r = \left(\frac{\partial\phi^r}{\partial\delta}\right)_\tau, \quad (3.3)$$

$$\phi_{\delta\delta} = \phi_{\delta\delta}^r = \left(\frac{\partial\phi}{\partial\delta^2}\right)_\tau, \quad \phi_{ss}^0 = 0, \quad \phi_{\tau\tau} = \phi_{\tau\tau}^0 + \phi_{\tau\tau}^0 = \left(\frac{\partial^2\phi}{\partial\tau^2}\right)_\delta,$$

$$\phi_{\tau\delta} = \phi_{\delta\tau} = \phi_{\delta\tau}^0 + \phi_{\delta\tau}^r = \left(\frac{\partial^2\phi}{\partial\tau\partial\delta}\right) = \left(\frac{\partial^2\phi}{\partial\delta\partial\tau}\right),...,\quad \text{etc.}$$

These quantities are directly obtained from an explicit form of the Equation (3.1) and simply defined the VCDs (Gorski, 1997; Gorski and Chmielniak, 1993):

$$z_{\delta\tau} \overset{def}{=} z + \delta\left(\frac{\partial z}{\partial \delta}\right)_\tau \equiv \delta(2\phi_\delta + \delta\phi_{\delta\delta}) = 1 + 2\delta\phi_\delta^r + \delta^2\phi_{\delta\delta}^r,$$

$$(3.4)$$

$$z_{\tau\delta} \overset{def}{=} z + \tau\left(\frac{\partial z}{\partial \tau}\right)_\delta \equiv \tau(\phi_\delta - \tau\phi_{\delta\tau}) = 1 + \delta\phi_\delta^r - \delta\tau\phi_{\delta\tau}^r,$$

Such nondimensional parameters (reduced to the values at thermodynamic critical point) are most useful in engineering practice and exhibit the discrepancies between real and ideal compressible fluid.

The auxiliary relations showing their physical significance are directly related to well-known coefficients of volume expansivity, β, isothermal compressibility, κ, and thermal expansivity, ε, of compressible fluid.

$$\left(\frac{\partial v}{\partial p}\right)_T\bigg/\left(\frac{\partial v}{\partial p}\right)_T^{id}=\frac{z}{z_{vT}}=\frac{z_{pT}}{z}, \quad \left(\frac{\partial p}{\partial T}\right)_v\bigg/\left(\frac{\partial p}{\partial T}\right)_v^{id}=\frac{z_{Tv}}{z}=\frac{z_{Tp}}{z_{pT}},$$

$$\left(\frac{\partial v}{\partial T}\right)_v\bigg/\left(\frac{\partial v}{\partial T}\right)_v^{id}=\frac{z_{Tv}}{z_{vT}}=\frac{z_{Tp}}{z}, \quad \text{where} \quad \beta=\frac{1}{v}\left(\frac{\partial v}{\partial T}\right)_p, \tag{3.5}$$

$$\kappa=-\frac{1}{v}\left(\frac{\partial v}{\partial p}\right)_T, \quad \varepsilon=\frac{1}{p}\left(\frac{\partial p}{\partial T}\right)_v, \quad \text{and} \quad \beta=p\cdot\kappa\cdot\varepsilon$$

The VCD parameters appearing in Equations (3.2) to (3.4) and Equation (3.6) only differ by the changing of state variables from absolute values to the reduced ones. In the case of reduced density $\delta=\rho/\rho_c$, and the reciprocal reduced temperature $\tau=T_c/T$, it is found that (Gorski, 1997; Gorski and Chmielniak, 1993):

$$z_{\delta\tau}\overset{def}{=}z+\delta\left(\frac{\partial z}{\partial\delta}\right)_\tau\equiv z+\rho\left(\frac{\partial z}{\partial\rho}\right)_T\equiv z_{vT}\overset{def}{=}z-v\left(\frac{\partial z}{\partial v}\right)_T;$$

$$z_{\tau\delta}\overset{def}{=}z-\tau\left(\frac{\partial z}{\partial\tau}\right)_\delta\equiv z_{Tv}\overset{def}{=}z+T\left(\frac{\partial z}{\partial T}\right)_v\equiv z+T\left(\frac{\partial z}{\partial T}\right)_\rho. \tag{3.6}$$

These auxiliary functions can be derived by the contact or Legendre transformation for any form of the EOS and a selected pair of two independent variables. In an explicit form of EOS, with respect to the pressure p (reduced pressure $\pi=p_r=p/p_c$), and the temperature T (as well as $\tau=T_c/T$), the proper pair of VCDs should be defined as:

$$z_{\pi\tau}\overset{def}{=}z-\pi\left(\frac{\partial z}{\partial\pi}\right)_\tau\equiv z_{pT}\overset{def}{=}z-p\left(\frac{\partial z}{\partial p}\right)_T;$$

$$z_{\tau\pi}\overset{def}{=}z-\tau\left(\frac{\partial z}{\partial\tau}\right)_\pi\equiv z_{Tp}\overset{def}{=}z+T\left(\frac{\partial z}{\partial T}\right)_p. \tag{3.7}$$

By eliminating the second subscripts in Equation (3.6), a shortened notation for the VCDs can sometimes be used but these functions are not independent (especially when distinguishing that $z_{Tp}\neq z_{Tv}$, and $z_{\tau\delta}\neq z_{\tau\pi}$). Ψ (T, p, V)

$$\Psi(T,p,\rho)=0 \quad \leftrightarrow \quad p=p(T,\rho), \tag{3.8}$$

and

$$\left(\frac{\partial T}{\partial p}\right)_\rho\left(\frac{\partial p}{\partial\rho}\right)_T\left(\frac{\partial\rho}{\partial T}\right)_p=-1.$$

From Equation (3.5) and Equation (3.8), one finds that

$$z_{Tv} > z > z_{vT}, \quad z_{Tp} > z_{pT} > z,$$

and

$$z_{pT} \cdot z_{vT} = z^2, \quad \frac{z_{Tv}}{z} = \frac{z_{Tp}}{z_{pT}}, \quad \frac{z_{Tv}}{z_{vT}} = \frac{z_{Tp}}{z},$$

$$z_{\pi\tau} \cdot z_{\delta\tau} = z^2, \quad \frac{z_{\tau\delta}}{z} = \frac{z_{\tau\pi}}{z_{\pi\tau}}, \quad \frac{z_{\tau\delta}}{z_{\delta\tau}} = \frac{z_{\tau\pi}}{z}. \tag{3.9}$$

The above given set of transfer relations allows one to obtain all important partial derivatives both from a particular thermal and fundamental EOS (Helmholtz free energy and Gibbs free enthalpy), and to transpose these results into a convenient form for practical use.

Based on these given relations, it is easy to find the proper forms for calculation of thermodynamic functions and such quantities as velocity of sound a_s, isentropic exponent k_s, Joule-Thomson coefficient μ_{JT}, Poisson ratio γ, and the Grüneisen parameter Φ. A few examples are presented in Table 3.1, (Gorski, 1997; Gorski and Chmielniak, 1993).

$$k_s = \frac{\rho}{p}\left(\frac{\partial p}{\partial \rho}\right)_s = -\frac{v}{p}\left(\frac{\partial p}{\partial v}\right)_s = -\gamma\frac{v}{p}\left(\frac{\partial p}{\partial v}\right)_T = \frac{c_p}{c_v}\frac{z_v}{z} = \frac{c_p}{c_v}\frac{z}{z_p}, \tag{3.10}$$

$$a_s \overset{\text{def}}{=} \sqrt{\left(\frac{\partial p}{\partial \rho}\right)_s} \equiv \sqrt{k_s p/\rho} = \sqrt{z_\rho \gamma RT} = z\sqrt{\gamma RT/z_p}, \tag{3.11}$$

$$\Phi = \frac{\rho}{T}\left(\frac{\partial T}{\partial \rho}\right)_s \equiv \frac{1}{\rho c_v}\left(\frac{\partial p}{\partial T}\right)_\rho = \frac{Rz_T}{c_v} = \frac{z_\rho}{z_T}\left(\frac{c_p}{c_v} - 1\right) = \frac{z_\rho}{z_T}(\gamma - 1), \tag{3.12}$$

where

$$z = z(T, \rho), \quad z_T \equiv z_{Tp} = z_{\tau\delta}, \quad z_\rho \equiv z_{pT} = z_{\delta\tau},$$

$$k_s = \frac{\rho}{p}\left(\frac{\partial p}{\partial \rho}\right)_s = -\frac{v}{p}\left(\frac{\partial p}{\partial v}\right)_s = -\gamma\frac{v}{p}\left(\frac{\partial p}{\partial v}\right)_T = \frac{c_p}{c_v}\frac{z}{z_p}, \tag{3.10'}$$

$$a_s \overset{\text{def}}{=} \sqrt{\left(\frac{\partial p}{\partial \rho}\right)_s} \equiv \sqrt{k_s p/\rho} = z\sqrt{\gamma RT/z_p}, \tag{3.11'}$$

where at $z = z(T, p)$, $z_T \equiv z_{Tp} = z_{\tau\pi}$, $z_p \equiv z_{pT} = z_{\pi\tau}$.

These identities express very simply both the important physical parameters and their discrepancies to an ideal gas model. The ultimate case shows simply that

$$\text{if } V = V(T,p) \Rightarrow \lim_{p \to 0}(z) = \lim_{p \to 0}(z_{Tp}) = \lim_{p \to 0}(z_{pT}) = 1,$$

$$p = p(T,V) \Rightarrow \lim_{V \to \infty}(z) = \lim_{V \to \infty}(z_{Tv}) = \lim_{V \to \infty}(z_{vT}) = 1 \qquad (3.13)$$

$$\text{and } a_s = \sqrt{\gamma R T}, \quad \Phi = (\gamma - 1), \quad k_s = \gamma = c_p/c_v,$$

Equation (3.13) and other above presented relations demonstrate what seems to be the simplest and most rational representation of discrepancies between real and ideal compressible fluid. In a formulation of VCDs, an additional positive can be observed — the transition of commonly used complicated mathematical relations not derivable from an EOS into an easily identified symbolic form; see Table 3.1.

Fundamental equations expressed in the form of the Helmholtz energy usually need temperature T and density ρ as input values to calculate thermodynamic properties. Due to the fact that in technical applications, different input values are given in most cases, the engineering programs contain iterations with which input values of the combinations (T, p), (T, h), (T, s), (p, ρ), (ρ, h), (ρ, s), (p, h), (p, s) and (h, s) can be handled in the homogeneous region as well as in the vapor-liquid two-phase region.

These iterations calculate the missing values of temperature T and density ρ with which all other properties can be calculated (Bejan, 2006). Extensive software packages have been developed during recent years (aspenTech, ChemicaLogic Corp., and Wagner and Overhoff, 2006), to make this modern way of easily calculating thermodynamic properties available for users in industry and research. These packages are customized for every individual field of application.

3.2 GAS MIXTURES

In the literature, a large number of fluid mixture models are available (Bejan, 2006, Orbey and Sandler, 1998, and Span, 2000). The models differ in both structure and accuracy. One group of models describes the behavior of mixtures through the use of excess properties. For instance, many models cited in Span (2000) have been developed for the excess Gibbs free enthalpy $G^{ex}(T, p, \overline{x})$, as well as for the excess Helmholtz free energy $A^{ex}(T, \rho, \overline{x})$.

To use these models, the pure components as well as the mixture itself must be in the same state at a given temperature and pressure (Orbey and Sandler, 1998). As a result of this precondition, these models are not suitable for most engineering problems when taking into account the involved components and the covered fluid regions. The thermodynamic properties of mixtures can be calculated in a very convenient way from the equations of state. Most of these equations are explicit in pressure, as well-established cubic equations of state. Cubic equations are still widely used in many technical applications due to their simple mathematical structure (Bejan, 2006; Orbey and Sandler, 1998). For technical applications with high demands on the accuracy of the calculated mixture properties, these equations show major weaknesses with respect to the representation of thermal properties in the liquid phase and the description of caloric properties.

Empirical equations of state, such as the equations of Bender (1973), yield an improved description of the properties of mixtures especially in the homogeneous region. These models are explicit in pressure as well. Bender used mixing rules to describe the composition dependence of the coefficients and the temperature-dependent functions of the equation of state. Starling (1973) used mixing rules for each coefficient of the equation of state (see Mäder 1991).

The other works laid the basis for the application of extended-corresponding-states models to mixtures (see Span, 2000). Based on this work, Sun and Ely (2005) developed the exact shape factor concept. Recently, Trusler et al., (in Span, 2000), reported two separate extended-corresponding-states models for natural gases and similar mixtures. One was a wide-ranging equation and the other was limited to the custody-transfer region.

The wide-ranging model yields an accuracy which is, on the average, similar to or even slightly better than other commonly used equations for natural gases. The mixture models mentioned above and the new model for natural gases and other technical gas mixtures are fundamental equations explicit in the Helmholtz free energy A with the independent mixture variables density ρ, temperature T, and molar composition $\bar{x}$.

Similar to fundamental equations for pure substances, the function $A(\rho,T,x)$ is split into a part A^0, which represents the properties of ideal gas mixture at a given density and temperature and the composition $\bar{x}$, and a part A^r, which takes into account the residual mixture behavior (Span and Wagner, 1996).

$$A(\delta,\tau,\bar{x}) = A^0(\delta,\tau,\bar{x}) + A^r(\delta,\tau,\bar{x}) \tag{3.14}$$

Using the Helmholtz free energy in its dimensionless form $\phi = A/(RT)$, Equation (3.14) reads

$$\phi(\delta,\tau,\bar{x}) \overset{def}{=} \frac{A(\delta,\tau,\bar{x})}{RT} = \phi^0(\delta,\tau,\bar{x}) + \phi^r(\delta,\tau,\bar{x}), \tag{3.15}$$

while δ is the reduced mixture density and τ is the inverse reduced mixture temperature according to

$$\delta = \rho/\rho_{ref} \quad \text{and} \quad \tau = T_{ref}/T, \tag{3.16}$$

with ρ_{ref} and T_{ref} being the reference composition-dependent reducing functions for the mixture density and temperature

$$\rho_{ref} = \rho_{ref}(\bar{x}), \quad T_{ref} = T_{ref}(\bar{x}) \tag{3.17}$$

The dimensionless form of the Helmholtz free energy for the ideal gas mixture ϕ^0 of the N components is given (Orbey and Sandler, 1998) as

$$\phi^0(\rho,T,\bar{x}) = \sum_{i=1}^{N} x_i \cdot \left[\phi_{0i}^0(\rho,T) + \ln x_i \right], \tag{3.18}$$

In the fluid mixture, ϕ_{0i}^0 is the dimensionless form of the Helmholtz free energy in an ideal gas state of component i and the x_i are the mole fractions of the mixture constituents. The term x_i in x_i accounts for the entropy growth by a mixing process.

In a multifluid approximation, the residual part of the reduced Helmholtz free energy of the mixture ϕ^r is given as follows:

$$\phi^r(\delta,\tau,\overline{x}) = \sum_{i=1}^{N} x_i \cdot \phi_{0i}^0(\delta,\tau,\overline{x}) + \Delta\phi^r(\delta,\tau,\overline{x}), \tag{3.19}$$

where ϕ_{0i}^0 is the residual part of the reduced Helmholtz free energy of the i-th component and $\Delta\phi^r$ is the departure function. The reduced residual Helmholtz free energy of each component depends on the reduced state variables δ and τ of the mixture; the departure function additionally depends on the mixture composition $\overline{x}$. This general structure corresponds to the models of Tillner-Roth (1993), Lemmon (1996), and Lemmon and Jacobsen (1999), as explained, respectively, by Span (2000), as well as by all models based on a multi-fluid approximation which intends to achieve an accurate description of the thermodynamic properties of nonideal mixtures (Span, 2000).

According to Equation (3.19), the residual part of the reduced Helmholtz free energy of the mixture ϕ^r is composed of two different parts (Span, 2000):

1. The linear combination of the residual parts of all considered mixture components
2. The departure function

In general, the contribution of the departure function to the reduced residual Helmholtz free energy of the mixture is inferior to the contribution of the equations for the pure components.

In summary, the fundamental equations of state for gas mixtures based on a multifluid approximation consists of:

- Pure substance equations for all considered mixture components
- Composition-dependent reducing functions $\rho_r = \rho_r(\overline{x})$, and $T_r = T_r(\overline{x})$ for the mixture density and temperature
- A departure function $\Delta\phi^r$ depending on the reduced mixture density, the inverse reduced mixture temperature, and the mixture composition

In an actual engineering practice, these models play a principal role for formulation of so-called standards for many fluid-state substances, both pure and compound (Span, 2000; Wagner and Kretzschmar, 2008). Part of the application of computer routines allows the introduction of such tools into the simulation of many advanced chemical and energy conversion processes. Some of the most popular software procedures are proposed by Wagner and Overhoff (2006) in the package "ThermoFluids" as well as IAPWS standard (Wagner and Kretzschmar, 2008), and other commercial software (Ruhr Universitat, 2007). As a part of a few very convenient online calculators, for example. NIST Chemistry WebBook (2007), two software packages, the ASPEN and Chemical Logic, play a leading role.

3.3 EFFICIENCY OF COMPRESSOR AND TURBINE FOR REAL GAS CONDITIONS

Real gas effects are extremely important when the compression and expansion processes take place within a high range of the pressure ratios. Such extreme conditions are typical while analyzing the supercritical Brayton and Rankine cycles. When supercritical CO_2 gas is expanded and compressed in a real gas fashion through a sequence of high-pressure turbines and compressors, the isentropic formulas shown in most thermodynamic books are no longer useful.

In order to obtain valuable results, it is necessary to develop more general procedures and equations for calculating the compressor and turbine efficiency in the dense gas region.

The objective of this analysis is to deliver mathematical derivations of the compressor and turbine isentropic and polytropic efficiency based on real gas models and frequently used thermal equations of state.

The real gas compression and expansion process has been analyzed by Kouremenos and Kakatsios (1985) and Wiederuch (1988), and is applicable as the standard VDI 2045-2:1993.

A more complete approach to this problem was proposed by Gorski (1997) based on an arbitrary form of thermal equation of state and VCDs explained in the previous chapter.

The commonly used term in analysis of compression and expansion processes is the polytropic efficiency. It refers to the elementary change of parameters at isentropic and polytropic processes and corresponding enthalpy increments; see Figure 3.1.

The polytropic efficiencies of compression η_{pc} and expansion η_{pe}, respectively, are:

$$\eta_{pc} = \left| \frac{dh_s}{dl_t} \right| = \left| \frac{v\,dp}{dh} \right|, \quad \eta_{pe} = \left| \frac{dl_i}{dh_s} \right| = \left| \frac{dh}{v\,dp} \right|, \tag{3.20}$$

where $dh_s = -v\,dp$ is an isentropic change of enthalpy for the adiabatic process, and specific technical work dl_t refers to the enthalpy rise dh in the real process.

The enthalpy changes with respect to the elementary pressure and volume increments are:

$$\left(\frac{dh}{dp} \right) = \left(\frac{\partial h}{\partial T} \right)_p \left(\frac{dT}{dp} \right) + \left(\frac{\partial h}{\partial p} \right)_T, \quad \left(\frac{dh}{dp} \right) = \left(\frac{\partial h}{\partial p} \right)_v + \left(\frac{\partial h}{\partial v} \right)_p \left(\frac{dv}{dp} \right) \tag{3.21}$$

With the first law of thermodynamics and Maxwell relations and the VCD, defined by Equation (3.6) and Equation (3.7), the mathematical derivation gives direct relations between temperature and pressure variation in an expansion polytropic process (Gorski, 1997):

$$\frac{dT}{T} - \frac{Rz}{c_p} \left(\eta_p + \frac{z_{Tv}}{z_{vT}} - 1 \right) \frac{dp}{p} = 0 \tag{3.22}$$

The similar relation for pressure and specific volume rise takes the form:

$$\frac{dp}{p} + \frac{k_s}{\left[1 - \dfrac{z_{vT}}{z_{Tv}}(\gamma-1)(\eta_p - 1)\right]} \frac{dv}{v} = 0 \, . \tag{3.23}$$

where $\eta_p \equiv \eta_{pe}$, and k_s is an isentropic exponent and γ is the specific heat ratio.

It can be pointed out that in the case of the compression process, the polytropic efficiency in Equation (3.22) and Equation (3.23) should be replaced by its reciprocal, i.e., $\eta_p \equiv 1/\eta_{pc}$.

The parameters appearing in Equation (3.22) and Equation (3.23) are not constants, but by an iterative procedure, can be averaged between initial 1 and final 2 states. Denoting their values as $\bar{\eta}_p, \bar{k}_s, \bar{z}, \bar{z}_{Tv}, \ldots$, these relations will be integrated and expressed in the typical form which is similar to an ideal gas polytrope.

An infinitesimal and integral counterpart of Equation (3.23) and Equation (3.24) can therefore be respectively written as:

$$\frac{dT}{T} - \bar{m}\frac{dp}{p} = 0, \quad \Leftrightarrow \quad p^{\bar{m}}/T = p_1^{\bar{m}}/T_1 = p_2^{\bar{m}}/T_2 = idem, \tag{3.24}$$

and

$$\frac{dp}{p} + \bar{n}\frac{dv}{v} = 0, \quad \Leftrightarrow \quad pv^{\bar{n}} = p_1 v_1^{\bar{n}} = p_2 v_2^{\bar{n}} = idem, \tag{3.25}$$

The averaged exponents $\bar{m}, \bar{n}$ for the compression of real gas are (Gorski, 1997):

$$\bar{m} = \frac{R}{\bar{c}_p \bar{\eta}_{pc}}\bigg|_{1,2} = \frac{\bar{\gamma}}{(\bar{\gamma}-1)\bar{\eta}_{pc}}\bigg|_{1,2}, \quad \bar{n} = \frac{1}{1-\bar{m}} = \frac{\bar{\gamma}}{1-(\bar{\gamma}-1)(1/\bar{\eta}_{pc}-1)}\bigg|_{1,2} \tag{3.26}$$

A more complete and symbolic representation of all basic thermodynamic processes in a dense gas is collected in Table 3.2.

TABLE 3.2
Generalized Exponents for Real Dense Gas Process

Process	$\bar{\eta}_{pc}, \bar{\eta}_{pe}$	$\bar{m}$	$\bar{n}$
$s = idem$	1	$R\bar{z}\bar{z}_{Tv}/(\bar{c}_p \bar{z}_{vT})$	$\bar{\gamma}\bar{z}_{vT}/\bar{z} = \bar{k}_s$
$h = idem$	$\approx \infty$	$(\bar{z}_{Tv}/\bar{z}_{vT} - 1)R\bar{z}/\bar{c}_p$	$\bar{k}_s/[1 + (\bar{\gamma}-1)\bar{z}_{vT}/\bar{z}_{Tv}]$
$T = idem$	—	0	$\bar{z}_{vT}/\bar{z}$
$p = idem$	0	∞	0
$v = idem$	—	$\bar{z}/\bar{z}_{Tv}$	∞

Source: Gorski, J. and Chemielniak, T., 1993.

It is clearly evident that in an ideal gas case, according to Equation (3.13), and other well-known principles of thermodynamics, all coefficients appearing in the above presented relations are constants and can be directly introduced into the integral forms of Equation (3.24) and Equation (3.25).

3.4 DETAILED SIMULATION OF A ZERO EMISSIONS POWER CYCLE ON PURE CARBON DIOXIDE

This section is aimed at describing a zero emissions quasi combined cycle, taking into account accurate data on CO_2 thermodynamic properties described in Section 3.1, and cooling the turbine blades and vanes. Oxygen supply is assumed from an external cryogenic air separation unit (ASU). Working substance is assumed as pure carbon dioxide. There are two principal problems with multistage compression and expansion of CO_2. First is the calculation of thermodynamic properties of a working medium and prediction of the cycle performance. Second is a detailed study of high-density flows in compressor/turbine stages and their influence on design criteria of turbomachinery (similarity parameters and mechanical constraints).

Discussed here are some thermodynamic aspects of CO_2 compression/expansion and a drop in a thermal efficiency caused by the imperfection of working fluid and associated losses (Yantovsky et al., 2002). In the modified arrangement of this semi-closed CO_2 cycle, see Figure 3.2, four large axial/centrifugal compressors (I/II/III/IVC) with cooling heat exchangers, two combustion chambers (CC1/CC2), ASU, oxygen compressor (OXC), regenerative heat exchanger (RHE), and three turbine stages (HPT/MPT/LPT) are considered. Carbon dioxide is cooled in the interstage

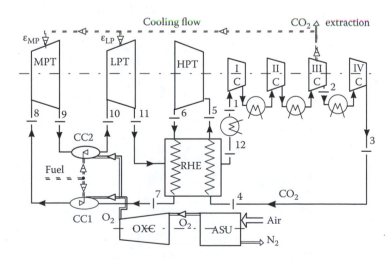

FIGURE 3.2 Scheme of modified zero emissions carbon dioxide supercritical cycle (Yantovsky et al., 2002). HPT/MPT/LPT = high/mean/low pressure turbine; I/II/III/IVC = CO_2 compressors; OXC = oxygen compressor; CC1, CC2 = combustion chambers; ASU = air separation unit; RHE = recovery heat exchanger. (From Yantovsky et al., 2002. With permission.)

heat exchangers up to an initial temperature $T_1 = 30°C$ after each stage compression (I, II and IIIC). In such cycle arrangement, a large amount of low-grade heat (300–500 kJ/kg at 140–160°C) can be extracted from the compressor coolers and used for a central district heating system.

The maximum pressure in the cycle $P_3 = 300$–310 bars will be achieved in the last IVC stage without cooling of the CO_2 stream because of a very small temperature growth (30–40°C). In our calculations, the amount of total energy consumption in the ASU equal to 780 kJ/kg O_2 is assumed. An expansion of high enthalpy CO_2 in LP turbine from $P_{10} = 10$ bars to $P_{11} \approx P_1$ allows the use of its energy for reheating of CO_2 before and after HP turbine in the RHE; see Figure 3.2. The back-pressure $P_9 = 40$–60 bars in MP turbine is regarded variable as well as the minimum pressure in the cycle $P_1 = 1$–5 bars.

In order to gain a good cycle performance, i.e., high thermal efficiency η_{th} and the net unit power $l_n = N_e/m$, the combustion of compressed fuel and oxygen in combustion chambers CC1 and CC2 takes place at the temperatures $T_8 = T_{10} = 1200$–1450°C and efficiency $\eta_b = 0.98$. A hypothetical pure carbon fuel (heating value: $Q_L = 30$ MJ/kg) has been assumed to avoid other components in the combustion products apart from carbon dioxide. Such a turbine inlet temperature (TIT) range needs to apply an advanced MPT/LPT cooling system and significantly affect the turbine characteristics. The turbine cooling streams ε_{MP}, ε_{LP} and the amount of CO_2 removed from the cycle (in percent of main stream) have been assumed as the bleed flow in compressor IIIC; see Figure 3.3. The corresponding TIT in HP turbine will be regarded as high

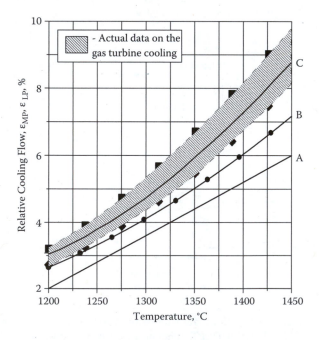

FIGURE 3.3 Turbine cooling flows. A—an optimistic variant, B—the most probable case, C—pessimistic variant. (From Yantovsky et al., 2002. With permission.)

as: $T_{HP} = 550–850°C$, depending on the operating conditions of RHE and its efficiency. The above conditions were essential in calculating the effectiveness of energy conversion processes in principal turbomachinery and other cycle components.

For an assumed level of polytropic/isentropic efficiency, pressure ratio, and mass flow in each compressor stage, it is possible to obtain the final parameters as well as the compression/expansion work and other characteristics based on a realistic model of CO_2 properties.

The simulation of CO_2 compression was performed by using virial equation of state (EOS) proposed by Altunin and Gadetskii (1971). An application of the VCD method (Gorski and Chmielniak, 1993; Gorski, 1997) to calculate CO_2 properties and compressor performance exhibits a large discrepancy between an ideal gas model and the real one. These differences are non-negligible in the prediction of energy consumption in the final two compressors, i.e., in the region of supercritical pressures (the error exceeds more than 20% in the compressor IIIC and 200% in the last IVC).

The total energy consumption of CO_2 compressors mainly depends on their efficiencies and an initial pressure P_1 in the cycle; see Table 3.3. An isentropic efficiency of the last (IVC) compressor is assumed $\eta_{IV} = 0.80$ and for other stages η_I, η_{II}, $\eta_{III} = 0.85$. The pressure losses in the pipelines and heat exchangers were taken into account (up to 2–3% of an inlet pressure). It is easy to show that the largest discrepancy between an ideal and real gas model of CO_2 properties can be observed at its final temperature and enthalpy in the outlet of each stage.

The tools and methods presented in Section 3.2 and Section 3.3 are essential for preparing a complete model of the cycle based on energy/mass flow balances for each element and the state parameters at the node points.

The supercritical zero emissions ion transport membrane oxygen power (ZEITMOP) cycle deals with very high pressures and temperatures. At these conditions, an ideal gas model, using simple isentropic compression and expansion cannot be applied because of real gas effects associated with nonideal gas compression and expansion processes.

TABLE 3.3
Comparison of an Ideal/Real Gas Work for CO_2 Compression

CO₂ Compression Work, l/[kJ/kg]		Initial Pressure of CO_2 in the Cycle, P_1 [bar]				
		1.0	2.0	3.0	4.0	5.0
Isentropic process[a]:	$l_s^{id} \rightarrow$	382.7	324.9	299.2	278.6	262.8
Polytropic process[a]:	$l^{id} \rightarrow$	456.7	394.0	358.4	334.1	315.6
Isentropic process[b]:	$l_s \rightarrow$	314.1	252.0	224.2	211.2	201.9
Polytropic process[b]:	$l_c \rightarrow$	371.5	298.4	266.0	250.6	239.7

[a] CO_2 – ideal gas, $c_p/c_v = 1.3$;
[b] CO_2 – real gas, Altunin's EOS (Altunin and Gadetskii, 1971).
Source: From Yantovsky et al., 2002. With permission.

Therefore, there was a need to develop analytical equations for polytropic expansion and compression for sequences of turbines and compressors, respectively.

In the following analysis, thermodynamic properties of CO_2 have been obtained from the virial EOS (Altunin and Gadetskii, 1971) and the data on specific heat c_p^{id} (T), given by:

$$z = z(T_r, \rho_r) = \frac{P_r z_c}{T_r \rho_r} = 1 + \sum_{i=1}^{6} \sum_{j=0}^{5} b_{ij} \cdot \rho_r^i \cdot T_r^{-j}, \quad \text{and} \quad c_p^{id} = \sum_{i=0}^{4} c_i T^i, \qquad (3.27)$$

where $P_r = P/P_c$, ... etc., b_{ij}, c_i—constants, P_c, T_c, ρ_c, z_c—parameters at the critical point.

The developed PC computer code allows finding all the additional and necessary data for CO_2. This code applies basic rules of calculation to the properties of gases (Walas, 1985) and the VCD method (Gorski, 1997) to improve them.

The conditions of turbine cooling correspond to the maximum TIT in the MPT and LPT turbines and limit both the net useful power and overall cycle performance. A consistent set of system studies reexamined this question in a reference to the earlier works (Yantovsky, 1995, and Mathieu, 1998). We examined three principal mode assumptions on the relation between turbine efficiency and relative cooling flows ε_{MP} and ε_{LP}: Case A (an optimistic one) corresponds to the approximate data on aero-gas turbines presented by Mattingly et al. (1987). An actual average set of these data serves to formulate Case B. The last one, Case C, gives rather a pessimistic prognosis for turbine efficiency decrement at high cooling flow requirements. These trends are presented in Figure 3.4 and Figure 3.5.

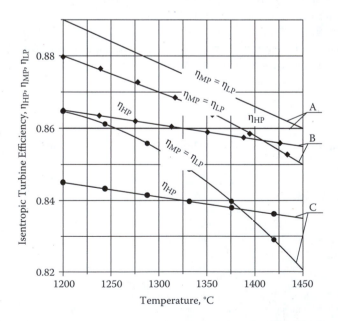

FIGURE 3.4 The assumed forms of HPT/MPT/LPT turbine efficiency variation.

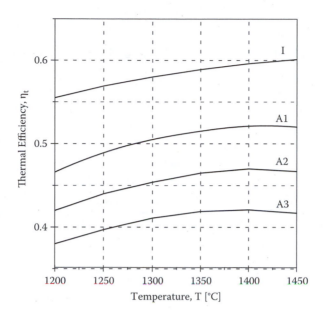

FIGURE 3.5 Obtained thermal efficiency of closed CO_2 cycle in comparison to the data (Yantovsky et al., 1995). I—previous data of Yantovsky et al. (1995); A1—data corresponding to an optimistic turbine cooling case: A and the pressure range: 1/10/40/300 [bar]; A2—as before but the pressures are: 2/10/50/300 [bar]; A3—the pressure range in the cycle: 3/10/60/300 [bar].

It is evident that, at the specified conditions in the basic node points and accompanying losses in the system components, the thermal efficiency of the cycle and normalized net turbine power should exhibit the optimum values with respect to TIT (T_8 and T_{10}). We were testing this tendency for previously obtained various conditions of the HPT/MPT/LPT turbine operation and additional results from simulation of compression processes as well as mentioned data on cooling effects.

Based on this information and the HPT/MPT/LPT efficiencies and cooling flows for Cases A, B, and C, the global quality coefficients of the cycle have been found. These results are exposed in Figure 3.5 through Figure 3.8, where the symbols: 1 to 6, and (I), (II), (III) correspond to the parameters presented in Tables 3.4 and 3.5. A comparison of obtained results and previous data presented by Yantovsky et al. (1995) and Mathieu (1998), shows that at comparable conditions, the overall cycle efficiency will be rather less optimistic; see Figure 3.5, curves I and A1. In Figure 3.6, symbols 1–6 exhibit the changes between thermal cycle efficiency η_{th} with respect to the net unit power l_n, kWs/kg.

It is clear (see Figure 3.5 through Figure 3.8) that optima thermal efficiency and net unit power in the cycle are close to the TIT equal to 1350–1450°C. It is the consequence of the blade cooling flow increase.

There is a great influence of the lowest pressure P_1 on the attainable cycle efficiency and useful net turbine power. Additionally, the particular values of cycle

TABLE 3.4
Data Set Utilized in the Cycle Simulation and Symbols Displayed in Figure 3.6

State Parameter Temperature TIT [°C]	Symbol					
	1	2	3	4	5	6
$T_{HP} \equiv T_5$	550	600	650	700	750	800
$T_{MP} \equiv T_8$	1,200	1,250	1,300	1,350	1,400	1,450
$T_{LP} \equiv T_{10}$	1,200	1,250	1,300	1,350	1,400	1,450

Turbine Pressure Ratio [bar]	Symbol		
	(I)	(II)	(III)
$\Pi_{HP} \equiv P_5/P_6$	300/40	300/50	300/60
$\Pi_{MP} \equiv P_8/P_9$	40/10	50/10	60/10
$\Pi_{LP} \equiv P_{10}/P_{11}$	10/1	10/2	10/3

performance parameters depend on the assumed level of turbine efficiencies and turbine cooling flow, as shown in Figure 3.5 through Figure 3.8.

The range of obtained values involves the necessity of searching for a compromise between thermodynamic performance of the cycle and some important fluid dynamics conditions. One can notice that it is better to chose $P_1 > 1$ bar in order to avoid (at the given mass flow) a large cross section of the first compressor stage as well as the last turbine stages. At the same time, we can maintain a low stress range in the rotating turbomachinery components and restrict the size of the recuperator.

The detailed study of zero emissions CO_2 supercritical cycle based on the real gas model showed a great influence of design parameters on the main components and cycle performance.

TABLE 3.5
Selected Data Used in the Cycle Simulation

State Parameter	Cycle Node Point								
	1	2	3	5	7	8	10	11	12
Temperature, °C	30	30	~65	550–850*	#	1250–1450*	1250–1450*	#	~80
Pressure, bar	1–3*	~84	210	~300	4–62	40–60*	10	1–3*	~1–3

Note: *range of assumed data, ~ approximate values, # unspecified before the calculation.
Source: From Yantovsky et al., 2002. With permission.

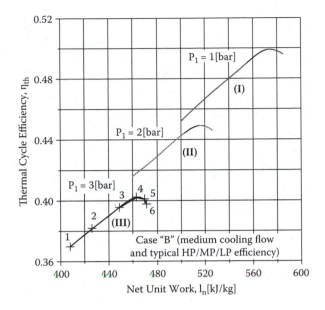

FIGURE 3.6 Thermal efficiency vs. net unit power of the cycle (Case B) (From Yantovsky et al., 2002. With permission).

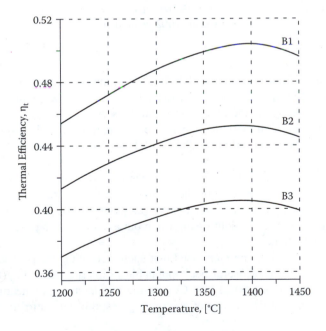

FIGURE 3.7 Thermal efficiency vs. maximum temperature of the cycle. B1—data corresponding to the pressure range: 1/10/40/300 [bar] and turbine cooling Case B; B2—the same cooling conditions, but pressure range: 2/10/50/300 [bar]; B3—as above, and pressure range: 3/10/60/300 [bar]. (From Yantovsky et al., 2002. With permission.)

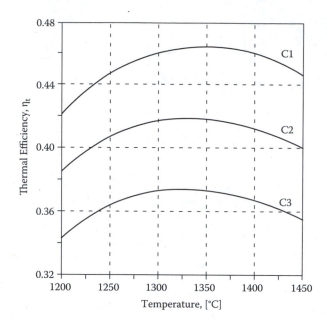

FIGURE 3.8 Thermal efficiency vs. maximum temperature of the cycle (Case C). C1—data are corresponding to the pressure range 1/10/40/300 [bar] and turbine cooling Case C; C2—as before, but the pressure range is: 2/10/50/300 [bar]; C3—cooling Case C and pressure range: 3/10/60/300 [bar]. (From Yantovsky et al., 2002. With permission.)

For a proper selection of the HPT/MPT/LPT turbine inlet conditions and the lowest pressure in the cycle, it is necessary to take into account the cooling streams and associated non-negligible thermal and hydraulic losses.

In order to get a high thermal efficiency and a great net useful power in a quasi-combined CO_2 cycle, the optimum TIT at MPT and LPT turbine should be 1350 to 1450°C.

The simplified model of combustion processes and assumed working medium (pure CO_2) is a good predictor of the performance of this zero emissions power cycles. An interesting indication of future developments is the coal gasification process by oxygen and CO_2 without steam, when our assumption on pure carbon fuel is a good approximation.

Furthermore, it should be mentioned that such a closed cycle is capable of supplying a great amount of useful heat applicable to district heating systems. Their estimated value is close 500 kJ/kg CO_2 at the temperatures of heating medium 130 to 150°C. In view of the lack of any stack gases, this is an attractive option for urban cogeneration.

An application of virial EOS and the VCD method allows for more effective compilation of the results of expansion/compression processes in high-pressure units. The results are comparable to the experimental data (a mean error in the prediction of thermodynamic properties not exceeding 0.8%).

REFERENCES

Airliquide Gas Encyclopedia: Gas Properties—General Data—Carbon Dioxide. http://ency-clopedia.airliquide.com/encyclopedia.asp?GasID=26#GeneralData, (accessed Jan. 15, 2008).

Altunin, V.V. and O.G. Gadetskii. 1971. An equation of state and thermodynamic properties of liquid and gaseous carbon dioxide. *Teploenergetica* 18(3): 81–83 (in Russian).

Altunin, V.V. and G.A. Spiridonov. 1973. "The system of experimental equations for calculating the thermodynamic properties of some important technical gases in a range from normal boiling temperature up to 1300°C and pressures 1000 bars," in *Issledovaniia po termodinamike*, ed. M. Vukalovich, Moskva: Nauka, 131–137 (in Russian).

AspenTech—Product Portfolio. www.aspentech.com/products/ (accessed Jan. 2, 2008).

Bejan, A. 2006. *Advanced Engineering Thermodynamics,* 3rd ed. NY: J. Wiley & Sons, Inc.

ChemicaLogic Corporation—CO_2Tab Product Information—Thermodynamic and Transport Properies of Carbon Dioxide, www.chemicalogic.com/co2tab/, (accessed Jan. 2, 2008).

Chiesa, P. and G. Lozza. 1998. "CO_2 abatement in IGCC power plants by semiclosed cycles," presented at *Int. Gas Turbine Congress*, Paper 98-GT-384, Stockholm, Sweden.

Gorski, J. 1997. *Modeling of real gas properties and its thermal-flow processes.* Rzeszow, Poland: Oficyna Wydawnicza Politechniki Rzeszowskiej (in Polish).

Gorski, J. and T. Chmielniak. 1993. "Prediction of real gas effects via generalized compressibility potentials," in *Proc. of Conf. ENSEC'93 on Energy Systems and Ecology.* July 5–9, Cracow, vol.1, 191–198.

Göttlicher, G. 1999. Energetik der Kohlendioxidrückhaltung in Kraftwerken, Fortschr.-Ber. VDI, Reihe 6, No. 421, Düsseldorf: VDI Verlag (in German).

Hangx, S.J.T. 2005. "Behaviour of the CO_2-H_2O system and preliminary mineralisation model and experiments." CATO Workpackage WP 4.1. Shell Int. Exploration & Production Program, (Shell contract No. 4600002284), 1–43.

Jericha, H. et al., 1998. Graz Cycle-eine Innovation zur CO_2 Minderung. *BWK*, Bd. 50 No.10 (in German).

Kerrick, D.M. and G.K. Jacobs. 1981. A modified Redlich-Kwong equation for H_2O, CO_2 and H_2O-CO_2 mixtures at elevated pressures and temperatures. *Am. J. Sci.* (281): 735–767.

Kouremenos, D.A. and X.K. Kakatsios. 1985. Ideal gas relations for the description of the real gas isentropic changes. Forschung im Ingenieurwesen, Bd. 51, No. 6: 169–174.

Li, H. Ji and X. Yan. 2006. A new modification on RK EOS for gaseous CO_2 and gaseous mixtures of CO_2 and H_2O. *Int. J. Energy Research* 30(3): 135–148.

Mäder, U.K. 1991. H_2O-CO_2, mixtures: A review of P-V-T-X data and an assessment from a phase equilibrium point of view. *Canadian Mineralogist* (29): 767–790.

Mathieu, P. 1998. Presentation of an Innovative Zero-Emission Cycle. *Int. J. Appl. Thermodynamics* 1(1-4): 21–30.

Mattingly, J.D. et al., 1987. *Aircraft Engine Design.* New York: AIAA Edu. Series.

NIST Chemistry WebBook—Thermophysical Properties of Fluid System; webbook.nist.gov/chemistry/fluid/ (accessed Nov. 15, 2007).

Orbey, H. and S. Sandler. 1998. *Modeling Vapor-Liquid Equilibria. Cubic Equations of State and Their Mixing Rules.* Cambridge: Cambridge University Press.

Ruhr Universitat of Bochum—Thermodynamics—Software / Equations of State; www.ruhr-uni-bochum.de/thermo/Software/Software_index-eng.htm /.

Sandler, S.I. 1999. *Chemical and Engineering Thermodynamics.* 3rd ed., New York: John Wiley & Sons.

Sengers, J.V., R.F. Kayser, and H.J. White, Jr. 2000. *Equations of State for Fluid and Fluid Mixtures.* Part I, IUPAC. Amsterdam: Elsevier Science B.V.

Span, R. 2000. *Multiparameter Equations of State.* Berlin/Heidelberg: Springer Verlag.

Span, R. and W. Wagner. 1996. A new equation of state for carbon dioxide covering the fluid region from triple-point to 1,100 K at pressures up to 800 MPa. *J. Phys. & Chem. Ref. Data* 25 (6): 1509–1597.

Starling, K.E. 1973. *Fluid Thermodynamic Properties for Light Petroleum Systems.* Houston: Gulf Publishing.

Sun, L. and J.F. Ely. 2005. A corresponding states model for generalized engineering equations of state. *Int. J. Thermophysics* 26(3): 707–728.

Van Wylen, G.J. and R.E. Sonntag. 1985. *Fundamentals of Classical Thermodynamics,* 3rd ed., New York: J. Wiley & Sons, Inc.

Vesovic, V. et al., 1990. The transport properties of carbon dioxide. *J. Phys. & Chem. Ref. Data* 19(3): 763–809.

Wagner, W. and H.J. Kretzschmar. 2008. *International Steam Tables—Properties of Water and Steam based on the Industrial Formulation IAPWS-IF97.* 2nd ed., Berlin/ Heidelberg: Springer Verlag.

Wagner, W. and U. Overhoff. 2006. *ThermoFluids. Interactive software for the calculation of thermodynamic properties for more than 60 substances,* (on CD). Berlin/ Heidelberg: Springer Verlag.

Walas, S.M. 1985. *Phase Equilibria in Chemical Engineering.* Vols. 1–2. New York: Buttersworth Publishing.

Wiederuch, E. 1988. An attempt to standardize the use of isentropic exponents for compressor calculations. *ASME Trans., Ser. A* (J. Eng. for Power), 110(2): 210–213. www.ruhr-uni-bochum.de/thermo/Forschung/Seiten/zustandsgleichungen-eng.htm (accessed Nov. 15, 2007).

Yantovsky, E. 1995. *Energy and Exergy Currents.* New York: NOVA Sci. Publ.

Yantovsky, E. et al., 1995. The cooperate—demo power cycle. *Energy Conv. Mgmt.* 36 (6–9): 861–864.

Yantovsky, E., Gorski, J., Smyth, B., and J. ten Elshof. 2002. "ZEITMOP Cycle (Zero Emission Ion Transport Membrane Oxygen Power)," in *Proc. of 15th Int. Conf. ECOS'02*, July 3–5, Berlin, vol. 2, 1153–1160.

4 Oxygen Ion Transport Membranes

4.1 NERNST EFFECT

In 1899, Walter Hermann Nernst observed the current of oxygen molecules through dense ceramics, when somewhat heated. The current of oxygen was similar to the current of electrons in metals under an electrical potential difference. The partial pressure of oxygen played the role of electrical potential. Some years later, he discussed this with Albert Einstein, and this resulted in the Nernst-Einstein formula:

$$j_{O_2} = \frac{\sigma_i RT}{4L(nF)^2} \ln\left(\frac{P'_{O_2}}{P''_{O_2}}\right). \tag{4.1}$$

Here j_{O_2} is the oxygen flux, F is Faraday's constant, L is the membrane thickness, n is the charge of the charge carrier ($n = 2$ for oxygen ions), R is the universal gas constant, T is the absolute temperature, P'_{O_2} is the oxygen partial pressure at the feed surface of the membrane, P''_{O_2} is the oxygen partial pressure at the permeate surface of the membrane, and σ_i represents the material conductivity. This expression clearly identifies the natural logarithm of the oxygen partial pressure ratio as the driving force for the oxygen flux.

An oxygen ion transport membrane (ITM) is a ceramic membrane made of one of the materials that conducts oxygen ions. They typically have perovskite or fluorite molecular structures, and contain oxygen ion vacancies, i.e., a *hole* in the molecular structure where an oxygen ion fits. When oxygen ions are excited, they can travel through the structure by leaping from vacancy to vacancy. As the membrane is a dense, impermeable ceramic, no gas can pass through, so the overall effect is one of a material that is permeable to oxygen and no other substance. Bouwmeester and Burggraaf (1997) explain the operation of an ITM:

> Dissociation and ionization of oxygen occurs at the oxide surface at the high pressure side (feed side) where electrons are picked up from accessible (near-) surface electronic states. The flux of oxygen ions is charge compensated by a simultaneous flux of electronic charge carriers. Upon arrival at the low-pressure side (permeate side), the individual oxygen ions part with their electrons and recombine again to form oxygen molecules, which are released at the permeate side.

To be certain, the ITM reactor description by R. Allam (2000) from Air Products is given:

> The ITM oxygen process uses nonporous, mixed-conducting, ceramic membranes that have both electronic and ionic conductivity when operating at high temperature, typically 800–900°C. The mixed conductors are inorganic mixed-metal oxides (e.g., perovskites such as (La,Sr)(Fe,Co,Cu)O3-d) that are stoichiometrically deficient of oxygen, which creates oxygen vacancies in their lattice structure. Oxygen from the air feed adsorbs onto the surface of membrane, where it dissociates and ionizes by electron transfer from the membrane. The resulting oxygen anions fill vacancies in the lattice structure and diffuse through the membrane under an oxygen chemical-potential gradient, applied by maintaining a difference in oxygen partial pressure on opposite sides of the membrane. At the permeate surface of the membrane, the oxygen ions release their electrons, recombine, and desorb from the surface as oxygen molecules. An electronic countercurrent accompanies the oxygen anion transport, eliminating any need for an external circuit. The separation is 100% selective for oxygen, in the absence of leaks, cracks, flaws, or connected through-porosity in the membrane.

Then, it is indicated that productivity of ITM is planned from 5 ton/day to 50 ton/day and commercialization is expected in the 2005 to 2007 time frame. Here, we stress the 50 ton/day O_2 capacity which is needed for a power unit of about 5 MW. (For recent data see Figure 7.16).

As the driving force is the partial pressure difference, the pure oxygen can be produced, as long as the total pressure on the permeate side is lower than the oxygen partial pressure on the feed side. If air is the feed gas, as is typically the case, this means the pressure on the permeate side must be less than 1/5 of the pressure on the feed side. By diluting the permeate, i.e., using a sweep gas, the oxygen partial pressure ratio can be increased without the need for a high total pressure differential across the membrane. If an oxygen-consuming reaction occurs at the permeate side, the oxygen partial pressure ratio is higher still. The question as to which is better, a *separate* membrane reactor to produce "artificial air" (oxygen with carbon dioxide or water vapor) and a *separate* combustion chamber, or *combined* air separation and combustion in the permeate side, is still to be determined. This question can only be answered by future tests. Some comparisons by computer simulation of both versions are given in Chapter 6.

One of the first papers to describe various schemes for adopting ion transport membranes for use in power production was Dyer et al. (2000). Oxygen production using membrane separation technology for gas-steam power production and internal gasifier integration were described. The authors have not used a sweep gas to remove the oxygen from the permeate side of the membrane, and instead used the pure oxygen in a coal gasifier. The resulting gas is combusted in air, and the exhaust is released into the atmosphere, so this is not a zero emission power plant (ZEPP).

ITM reactors have many design problems, but stability is a crucial one. Oxygen flux is inversely proportional to the thickness and manufacturers are currently making membranes of the order of tens of micrometers thick. These thin dense membranes must be supported on a porous substrate, particularly if there is a difference in pressure ratio across the membrane. The porous substrates may be made of the same or similar material to the membrane, i.e., ceramic. But the reactor operates at high temperatures, which can cause porous ceramic to sinter, reducing the porosity and hindering performance. Van der Haar (2001) reported that:

> Mechanical tests of the porous perovskite support reveals that these could endure an absolute pressure difference of about 30 bar ... application of the supports at temperatures close to 1000°C will reduce its porosity due to non-negligible sinter activity at these temperatures.

These mechanical problems are currently being addressed by a number of companies and research groups around the world.

We would like to present technical information about currently available ITM reactors. However, the main characteristic of any reactor which is crucial for its size and cost, is the achieved oxygen flux $\dot{J}_{O_2}$. No manufacturers are currently publishing this information. Instead, they are giving information such as the relative increase in oxygen flux during development, or describing the overall size of a reactor in general terms.

Foy and McGovern (2005) compared published data from laboratory tests for a number of different ITM materials. Most of these tests were performed on relatively thick samples on the order of 1 mm. As mentioned above, manufacturers are currently working with thicknesses of tens of micrometers. Table 4.1 shows the materials compared, and Figure 4.1 and Figure 4.2 show the results of the comparison. P_1 is the oxygen partial pressure on the feed side and L is the thickness. The unit of flux is 1 µmole/cm$^2 \times$ s [= 0.32 g/m$^2 \times$ s].

Lawrence Livermore National Laboratory developed a global description of the oxygen permeation through dense ceramic membrane (Pham, 2002). This theory predicted that:

> Oxygen flux as high as 100 ml per sq.cm × min is possible if the surface of the membrane is coated with a high surface area catalyst layer.

TABLE 4.1
ITM Materials Compared by Foy and McGovern (2005)

Name	Formula	Author	Year
BBCF	$BaBi_{0.4}Co_{0.2}Fe_{0.4}O_{3-d}$	Shao et al.	2000
BCF	$BaCe_{0.15}Fe_{0.85}O_{3-d}$	Zhu et al.	2004
BSCF	$Ba_{0.5}Sr_{0.5}Co_{0.8}Fe_{0.2}O_{3-d}$	Wang et al.	2002
BTCF	$BaTi_{0.2}Co_{0.5}Fe_{0.3}O_{3-d}$	Tong et al.	2003
CLFC	$Ca_{0.6}La_{0.4}Fe_{0.75}Co_{0.25}O_{3-d}$	Diethelm et al.	2003
LCF	$La_{0.4}Ca_{0.6}FeO_{3-d}$	Diethelm et al.	2003
LCFC	$La_{0.6}Ca_{0.4}Fe_{0.75}Co_{0.25}O_{3-d}$	Diethelm et al.	2004
LSC	$La_{0.5}Sr_{0.5}CoO_{3-d}$	Van der Haar	2001
LSCF	$La_{0.6}Sr_{0.4}Co_{0.2}Fe_{0.8}O_{3-d}$	Shao	2003
LSGF	$La_{0.15}Sr_{0.85}Ga_{0.3}Fe_{0.7}O_{3-d}$	Shao	2003
LSGF-BSCF	$La_{0.15}Sr_{0.85}Ga_{0.3}Fe_{0.7}O_{3-d} Ba_{0.5}Sr_{0.5}Fe_{0.2}Co_{0.8}Fe_{0.2}O_{3-d}$	Wang et al.	2003

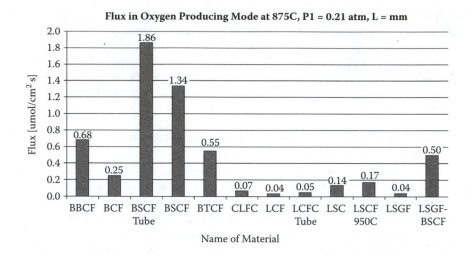

FIGURE 4.1 Normalized flux for the various materials (Foy and McGovern, 2005), umol = µmole.

The limit mentioned here is about 23 g/m²s. In contemporary design, many authors assume a membrane flux of 1 g/m²s (3.125 µmole/m²s).

4.2 OXYGEN ION TRANSPORT MEMBRANE REACTORS FOR ZEPPS

The largest element of the AZEP cycle is the membrane reactor. This large ceramic module operates at temperatures of 1250°C. Air heaters for coal powder-fired air turbines are another example of large ceramic bodies at high temperatures. After many

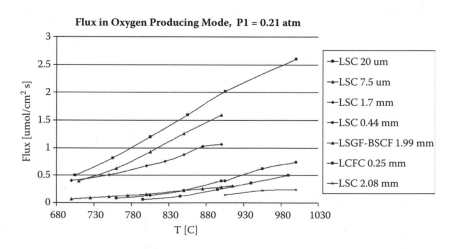

FIGURE 4.2 Actual non-normalized fluxes for membranes with different physical thicknesses. The names are listed in order of decreasing flux (Foy and McGovern, 2005). Here, umol = µmole.

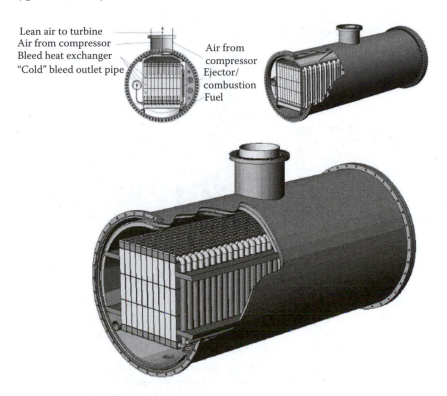

Lean air to turbine
Air from compressor
Bleed heat exchanger
"Cold" bleed outlet pipe

Air from
compressor
Ejector/
combustion
Fuel

FIGURE 4.3 Design of ITM reactor for AZEP cycle (Sundkvist et al., 2004).

decades of work, these air heaters are still a problem. It is possible that development of an integrated OITM/combustion chamber might be more difficult than development of a CO_2 turbine. The current design of the module is shown in Figure 4.3. Combustion chambers and ITM modules are incorporated into the same chamber.

The reactor temperature is controlled by the temperature of combustion. The fragility of the materials under consideration means that the temperature of the combustion must be relatively low (less than 1250°C). Such a low temperature results in a relatively low efficiency. In addition to this, the AZEP group has identified staged combustion using partial catalytic oxidation as the optimum method of achieving low temperature complete combustion (Sundkvist et al., 2004). This complicated method of combustion brings its own engineering challenges.

The ITM modules are based on an extruded ceramic structure shown in Figure 4.4. MCM, or mixed conducting membrane, is the ITM material.

The extruded ceramic is a porous support, which is coated with a dense membrane, as shown in Figure 4.5.

Further development of the AZEP reactor is reported by Selimovich (2005). This paper gives comprehensive information on many aspects of the reactor development. Many of the engineering challenges for the reactor are similar to those faced by heat exchanger designers, for example, improving the surface to volume ratio. Dealing

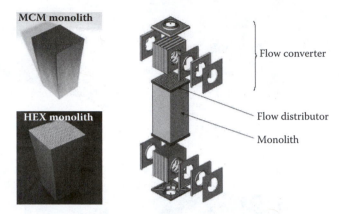

FIGURE 4.4 Design of ITM reactor for AZEP cycle (Sundkvist et al., 2004).

with nonuniform flow distribution is a headache for all designers of chemical and nuclear reactors, and the AZEP reactor is no exception. Selimovich (2005) gives detailed information on the various options under consideration for the solution of these problems. He identifies counterflow as more effective than co-flow, which is in accordance with heat exchanger theory, and also presents data on ITM materials. Table 4.2 shows comprehensive data on the reactor development. Note the oxygen partial pressure at the inlet is 20.7 kPa, implying that the inlet air is at atmospheric pressure. The operation pressure is defined as 10 bars. Tests have shown that the system operates as expected to a pressure of 10 bars and a temperature of 900°C.

In addition to the engineering challenges inherent in the design of the unit, the reactor also has high maintenance costs. It seems likely that the ceramic parts will have a life of 2.5 to 7.5 years.

FIGURE 4.5 Porous support with dense membrane (Sundkvist et al., 2004).

TABLE 4.2
Technical Data for AZEP Reactor

Temperature of inlet air (MCM)	700°C
Temperature of inlet sweep (MCM)	1000°C
Operation pressure	10 bars
Oxygen partial pressure (inlet air side)	20.7 kPa
Oxygen partial pressure (inlet seep side)	0.8 kPa
Hydraulic diameter (square channel)	2 mm
Wall thickness	0.6 mm
MCM length	0.4
Number of repeating units	500
Porosity (porous support)	0.32
Tortuosity factor (porous support)	2.2

Source: Selimovich, F., 2005.

Renz et al. (2005) presents detailed information on the design of the membrane reactor for the Oxycoal-AC cycle. Mechanical stability of the ceramic is again provided by using dense membranes on porous supports. Two shapes are compared—tubular membranes with cross-flow and planar membranes with counter-flow, as shown in Figure 4.6 and Figure 4.7. Detailed information on pressures and temperatures in the unit is presented, along with calculations showing stress in the ceramic (Renz et al., 2005).

In recent years, a small (1 m² of an active surface) prototype of ITM reactor has been designed and tested. It is made of BSCF ceramic tubes in the joint work of RWTH Aachen; see Figure 4.8. The measured and published data on oxygen flux are more representative than measured on a small, coin-sized samples.

The tests (Figure 4.9) vividly show the oxygen flux versus temperature by different pressures and unsatisfying tolerance to CO_2.

BSCF ceramic is an oxide of barium, strontium, cobalt, and ferrum; see Figure 4.1 above. It is possible to compare the data of RWTH with the upper line of Figure 4.2 which reflect other measurements in the same ceramic tubes. In most engineering

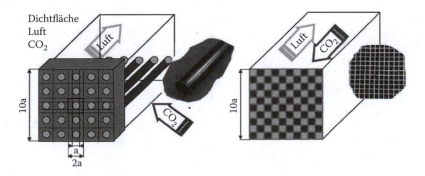

FIGURE 4.6 Possible structures for membrane rector in Oxycoal-AC cycle (Renz et al., 2005). Dichtfläche = sealing surface; Luft = air.

FIGURE 4.7 Membrane tested for Oxycoal cycle (Renz, 2004).

projects, the unit of flux is 1 g/m^2 × s (gram per square meter and second). The permeation unit in Figure 4.9 is 1 ml/cm^2 × min = 0.238 g/m^2 × s, whereas in Figure 4.2, it is 1 μmole/cm^2 × s = 0.32 g/m^2 × s. In the lower point by 650°C, we have 0.238 × 0.5 = 0.119 and 0.32 × 0.5 = 0.16 g/m^2 × s. In the upper point by 900°C, we have 0.64

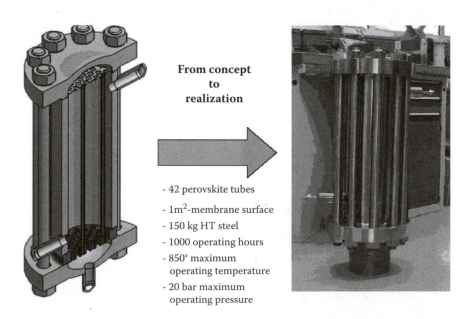

From concept
to
realization

- 42 perovskite tubes
- 1m^2-membrane surface
- 150 kg HT steel
- 1000 operating hours
- 850° maximum operating temperature
- 20 bar maximum operating pressure

FIGURE 4.8 ITM prototype reactor made of BSCF ceramic tubes (RWTH Aachen).

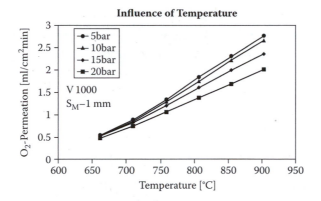

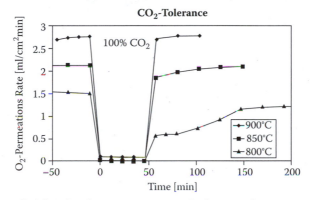

➤ Slightly reduced permeation compared to literature data

➤ Reason: tube geometry D_{Mem} = 15 mm, dilution resp. enrichment towards the membrane surface

➤ Unsatisfying CO_2-tolerance

FIGURE 4.9 Tests of BCFC ceramics at different operating conditions (RWTH Aachen).

and 0.32 g/m^2 × s. The agreement is quite good. It supports the trust in measurement by coin-sized samples.

4.3 CHEMICAL LOOPING COMBUSTION

Knoche and Richter (1968) proposed using metal as a carrier for oxygen to increase combustion efficiency. Metals would be oxidized using air, and the metal oxides would then be reduced by a fuel in a separate chamber. Although the aim of Knoche and Richter was to increase combustion efficiency, this method of combustion, now called "chemical looping combustion," provides a means of zero emissions combustion.

Ishida developed and experimentally proved this concept for zero emissions (Ishida and Jin, 1998). A number of research groups are investigating chemical

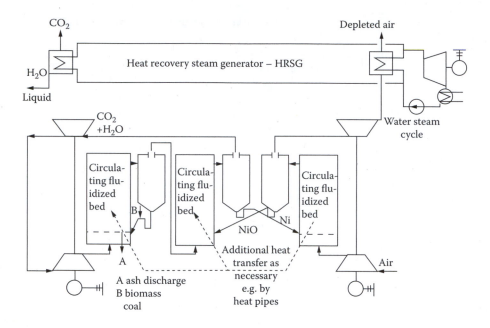

FIGURE 4.10 Coal or biomass fired zero emissions cycle using metal as the oxygen carrier (Leithner et al., 2005).

looping combustion and it is currently at the laboratory scale. Leithner et al., (2005) show a schematic, reprinted here as Figure 4.10, for a coal or biomass fired cycle using nickel as the oxygen carrier.

Leithner notes that this concept is similar to that of oxygen transport using oxygen ion transport ceramics. He demonstrates this with a similar schematic using a membrane instead of the metal oxide subsystem, shown in Figure 4.11. This cycle concept is the same as the Milano cycle (Romano et al., 2005) or the Oxycoal cycle (Renz et al., 2005). The use of ceramic membranes instead of metal oxides removes the need for two of the circulating fluidized bed reactors. Use of membranes is currently at a more advanced stage than chemical looping combustion. Tortuosity, described as the increase in combustion efficiency using chemical looping which sufficiently compensates for the greater mechanical complexity of the cycle, remains to be seen.

There is some commercial interest in chemical looping combustion. The U.S. Department of Energy has granted funding to a group, who are using flue gas recycling to burn coal in a mixture of oxygen and flue gas (Anna, 2005).

Perovskite is an oxygen ion ceramic used in ion transport membranes, but it seems that BOCs are using perovskite as the oxygen carrier in a chemical looping system. Use of ion transport ceramics instead of metal oxides in chemical looping is an interesting new development.

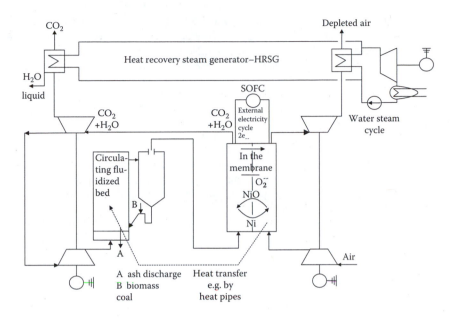

FIGURE 4.11 Coal or biomass fired zero emissions cycle using ion transport membrane as the oxygen carrier (Leithner et al., 2005).

REFERENCES

Allam, R.J. 2000. "Air Separation Units, Design and Future Development," in *Proc. of ECOS'02*, University of Twente, July 5–7, Enschede, the Netherlands.

Anna, D. 2005. "DOE Advances Oxycombustion for Carbon Management." DOE/NETL, www.fe.doe.gov/news/techlines/2005/tl_oxycombustion_award.html (accessed Oct. 16, 2005).

Armstrong, P., J. Sorensen, and T. Foster. 2003. JTM oxygen: an enabler for IGCC. *Gasification Technol*. 12–15 October.

Bouwmeester, H.J.M. and A.J. Burggraaf. 1997. Dense ceramic membranes for oxygen separation, in: *The CRC Handbook of Solid State Electrochemistry*. Gellings, P. J. and H.J.M. Bouwmeester (eds.), Boca Raton, FL: CRC Press, 481–552.

Dyer P. et al., 2000. Ion transport membrane technology for oxygen separation and syngas production. *Solid State Ionics* (134): 21–33.

Foy K. and J. McGovern. 2005. "Comparison of Ion Transport Membranes," in Proc. 4th Annual Conference on Carbon Capture and Sequestration. May 2–5, Alexandria, VA, U.S.A., Paper 111.

Ishida, M. and H. Jin. 1998. "Greenhouse gas control by a novel combustion: no separation equipment and energy penalty," presented at 4th Int. Cong. GHGT, ENT-13, Interlaken, Switzerland.

Knoche, K.F. and H. Richter. 1968. "Verbesserung der Reversibilitaet von Verbrennungsprozessen," *Brennstoff-Wärme-Kraft* (BWK) 20 (5): 205 (in German).

Leithner, R. et al., 2005. Energy conversion processes with intrinsic CO$_2$ separation, *Trans. of the Society for Mining, Metallurgy and Exploration* (318): 161–165.

Pham, A.Q. 2002. "High-efficiency steam electrolyzer," in Proceedings of the 2002 U.S. DOE
      Hydrogen Program Review NREL/CP-610-32405. www1.eere.energy.gov/ hydroge-
      nandfuelcells/pdfs/32405b26.pdf (accessed Nov. 27, 2006).
Renz U. et al., 2005. "Entwicklung eines CO2-emissionsfreien Kohleverbrennungs-processes
      zur Stromerzeugung," CCS-Tagung Juelich, Nov. 10–11, RWTH, (in German).
Romano M. et al., 2005. "Decarbonized Electricity Production from Coal by means of Oxygen
      Transport Membranes," in 4th Annual Conference on Carbon Sequestration, May 2–5,
      Alexandria, VA, 1402–1412.
Selimovich, F. 2005. "Modeling of Transport Phenomena in Monolithic Structures related
      to CO_2-Emission Free Power," ISRN LUTMDN/TMHP—05/7029-SE. Thesis for
      degree of Licentiate in Engng. Dept Heat & Power Eng., Lund Inst. of Technology.
Sundkvist, S. et al., 2004. "AZEP gas turbine combined cycle power plants," in GHGT-7,
      University Regina, Canada. http://uregina.ca/ghgt7/PGF/papers/peer/079.pdf (accessed
      Apr. 9, 2006).
van der Haar, M. 2001. Mixed conducting perovskite membranes for oxygen separation. Ph.D.
      thesis, University of Twente, Enschede, the Netherlands.

5 The ZEITMOP Cycle and Its Variants

5.1 THE ZEITMOP CYCLE WITH SEPARATE ITMR AND COAL-POWDER FIRING

The schematic of the power cycle is presented in Figure 5.1 and is similar to the gas-fired version; see Figure 2.14 in Chapter 2. Ambient air enters compressor 1 which, following compression, is heated up to approximately 800 to 900°C in heat exchanger 3 by the flue gases exiting turbine 6. The hot pressurized air then enters the ITM oxygen ceramic, 4, which separates the air into a relatively high pressure oxygen-depleted air stream and pure oxygen, the latter penetrating the membrane and mixing with the CO_2 flow exiting turbine 9. The flow of swept oxygen and CO_2 forms an oxidizer for the fuel gas entering combustor 7. The hot pressurized oxygen-depleted air stream leaving reactor 4 is expanded in turbine 5 before being discharged to the atmosphere. The flue gas mixture of CO_2 and H_2O exiting combustion chamber 7 (at approximately 1300–1600°C) is expanded in low pressure turbine, 6, before being cooled in exchanger 3, recuperator 10, and cooling tower 13. At ambient temperature, the water in the flue gas mixture is in liquid form while the CO_2 remains gaseous. The bulk of the water, therefore, is deflected out of the cycle in 12. Almost pure CO_2 enters the multistaged inter-cooled compressor, 11, where a fraction of highly compressed (200–300 bars), supercritical or liquid CO_2 is deflected out of the cycle in 14 to be sequestered. The major portion of the CO_2 is heated in 10 before being expanded in the high pressure turbine, 9, down to approximately 15 to 40 bars. The CO_2 then enters the permeate side of 4 to sweep oxygen.

In order to protect turbine 6 and the heat exchangers from erosion, a system for removing liquid ash and alkali is employed, proven by Förster et al. (2001) (VGB Power Technology 2001, No. 9). A similar system could be used for the combustion of residual oil.

The ITM ASU should not deviate from the projected performance of Air Products. The company aims to produce 50 t O_2/day in the near future. Below is described a membrane separation unit in a new zero emissions power cycles, which has never been outlined by Air Products.

Some results of the calculations for a separate ITM ASU are as follows (Yantovsky et al., 2003a):

> Membrane thickness, 20 micrometers; electrical conductivity, 1/ohm × m; temperature, 1193 K; O_2 production rate, 4 kg/s; air mass flow rate, 25 kg/s; air pressure, 25 bars; O_2 pressure in feed, 5.25 bars; N_2 pressure in feed, 19.76 bars.

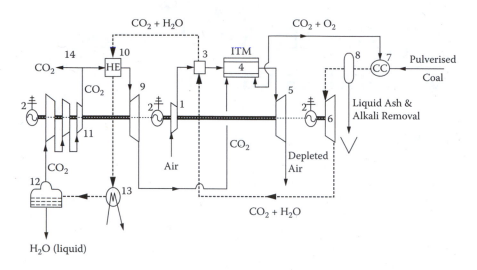

FIGURE 5.1 ZEITMOP power cycle (using *pulverized coal*) with separate combustion and ITM reactor. 1-air compressor, 2-synchronous electrical machine, 3-heat exchanger (recuperator), 4-ITM reactor, 5-depleted air turbine, 6-CO_2 + H_2O turbine, 7-combustor, 8-liquid ash and alkali removal, 9-CO_2 turbine, 10-recuperator, 11-CO_2 compressor, 12-water separator, 13-cooling tower, 14-CO_2 release.

> Inside reactor on air side: mass flow rate, 21 kg/s; total pressure 21.4 bars; N_2 pressure, 19.76 bars; O_2 pressure, 1.63 bars; percent of O_2 depletion, 68.8%.
>
> Inside reactor on permeate side: O_2 pressure, 1 bar; ratio of O_2 partial pressures, 1.63; (therefore: $\ln 1.63 = 0.49$); O_2 flux = 0.0522 g/m^2 × s; surface area required, 76,700 m^2; volume requirement 7.2 × 7.2 × 7.2 m^3 = 373.2 m^3.

5.2 GAS-FIRED ZEITMOP VERSION WITH COMBINED ITMR AND COMBUSTOR

Apart from pure O_2 production, the ITM unit is also used in industry for syngas production according to the chemical reaction:

$$CH_4 + \frac{1}{2}O_2 \Rightarrow CO + 2H_2 \qquad (5.1)$$

Syngas, CO and H_2, are saleable products, the raw material for numerous chemical syntheses. Its production is examined in detail by Ritchie et al. (2004). The most general analysis is made by Korobitsyn et al. (2000) in ECN. A reactor operating temperature of 1121°C and an operating pressure of 27 bars, for a methane conversion rate higher than 95%, were assumed in the calculations. The oxygen flux through the membrane was 10 Nml/cm² × min = 2.3 gO₂/m² × s. If such a flux were to be achieved in the above mentioned ITM separation unit, its dimensions would be $(2.3/0.0522)^{1/3} = 3.47$ less for each dimension, i.e., about $2.1 \times 2.1 \times 2.1$ m³ $= 9.26$ m³.

The essence of this application is a further step in the oxidation of methane: its complete oxidation:

$$CH_4 + 2O_2 \Rightarrow CO_2 + 2H_2O \tag{5.2}$$

In contrast to the reaction in Equation 5.1, which is an endothermic reaction, Equation 5.2 is exothermic. It is a well-known reaction for the combustion of methane. In contrast to the ordinary combustion chambers of gas turbines or piston engines, the use of an ITM unit prevents the combustion products, $CO_2 + H_2O$, from being diluted with nitrogen. At ambient temperature, water in liquid form can easily be separated from the gaseous CO_2. As the latter forms the working fluid of the quasi-combined zero emissions cycle, the deflection of a fraction (5–10%), of this highly compressed gas flow for subsequent sequestration, does not present any difficulty.

For the selection of the ceramic membrane material and oxygen flux evaluations, the detailed research results of ten Elshof (1997) are used. For the ITM combustor (i.e., reactor) configuration, the well-documented shell-and-tube configuration is used; see Figure 5.2. The tubes are made of a porous refractory ceramic such as alumina, Al_2O_3. The mixed conducting ceramic membrane is sprayed onto the porous substrate. Many such thin film deposition technologies are known and more are under development. The thickness of the porous alumina wall is approximately 2 to 3 mm, while the thin-film membrane layer is on the order of tens of micrometers. The proposed scheme of a zero emissions power plant with an ITM combustor is presented in Figure 5.3.

1=ITM ceramic tube
2=cupweld of steel & ceramic (argon arc weld)
3=steel tube disc
4=end cap
5=steel shell (stainless steel)
6=clearance gap
7=thermal insulation (quartz)

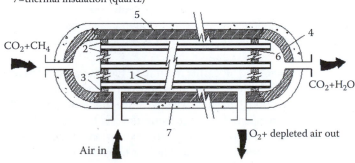

FIGURE 5.2 ITM reactor-combustor (similar to Figure 4.8).

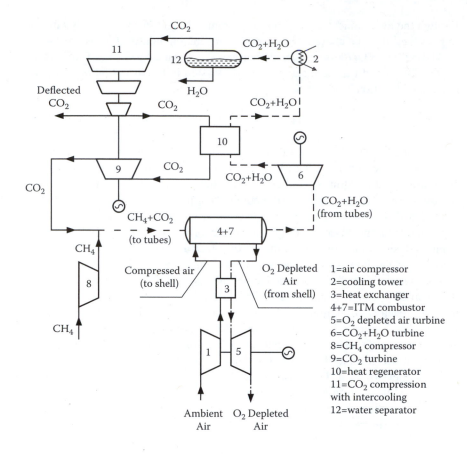

FIGURE 5.3 The ZEITMOP cycle with combined ITM reactor and combustor.

Air enters compressor 1 and heat exchanger 3 and, after reaching the required pressure and temperature, enters the shell (feed) side of combustor 4 + 7. After losing some oxygen and being heated to a higher temperature due to the heat supply from the tube side, the oxygen-depleted air flows through exchanger 3 and air turbine 5, driving the air compressor. The O_2 depleted air is released into the atmosphere without any harm. It is free of any contaminant.

Inside the membrane-coated tubes, the mixture of fuel gas from 8 and re-circulated CO_2, enters from the left, both at the required pressures and temperatures to absorb the oxygen flux passing through the tube wall. Near the tube wall surface, the diffused oxygen reacts with the methane fuel, increasing the gas temperature. On the right-hand side of the ITM combustor, mirroring the exit of an ordinary combustor, the combustion products, CO_2 and H_2O, are at a higher temperature than at the entrance. The chemical energy of the methane is converted into the thermal (internal) energy of the combustion products. The combustion products are then expanded in turbine 6 and give up heat to the recuperator, 10, before being cooled in cooling tower 13, allowing water to be separated from the CO_2 in 12. Almost pure CO_2 is compressed in a multistage inter-cooled compressor, 11. The bulk portion is then heated in recuperator 10 and expanded in turbine 9, the latter driving the CO_2 compressor. After expansion, the CO_2 is at the required pressure and temperature so it can mix with the fuel gas and enter the combustor. The electrical devices on all three shafts are synchronous generators, which may work as motors in starting and transient regimes.

5.3 A ZERO EMISSIONS BOILER HOUSE FOR HEATING AND COOLING

The diversity of the energy supply required in Europe is dominated by heating, not power (i.e., electricity.) The scheme shown in Figure 5.3 can easily be converted for the co-generation of heat and power. In practice, however, boiler houses alone produce the required heat. Boiler houses are responsible for a highly significant portion of greenhouse gas emissions. A zero emissions boiler house is outlined in Figure 5.4. Such a boiler house can produce cold air as well and uses the same ITM combustor as shown in Figure 5.2. It works as follows.

Ambient air enters compressor 1, and is heated in exchanger 3 before entering the feed side of the ITM combustor, 4. After flowing along the shell side (i.e., the outer tubes) of the ITM combustor, the air loses about 70% of its oxygen content. It is then expanded in turbine 5 before being discharged to the atmosphere without causing any environmental damage. Compressed fuel gas from 8 is mixed with pure (recycled) CO_2 before entering the permeate side (i.e., the inner tubes) of combustor, 4. The temperature of the permeate side gases increases as a result of the combustion which takes place due to the movement of O_2 through the tubes from the shell side. The hot gases are first cooled in recuperator, 10, before giving up heat to the hot water system in 14 for space heating and then the hot water supply system, 13. The temperature of the water entering the hot water system is ambient temperature (i.e., approximately 10–15°C). The combustion products of CO_2 and H_2O are at a pressure of about 5 to 20 bars. Water is deflected in 12, while the gaseous CO_2 is compressed to at least 70 bars in compressor 11. Upon exiting compressor 11, a fraction of the CO_2 is transferred to be liquified and sequestered while the bulk returns via recuperator 10 and turbine 9 to be mixed with the fuel before entering the tube side of the ITM combustor 4, thus completing the cycle.

If local cooling is required, some air exiting 1 can be deflected to the air cooler 6 where it is cooled to ambient temperature. This air is then expanded in combustor 7, lowering the air temperature for cooling purposes.

5.4 A TRANSPORT POWER UNIT VERSION USING A TURBINE

This section discusses only turbine units, leaving piston engines to be considered in Chapter 7. The emissions from transport engines, especially cars, dominate the pollutant mix, particularly in urban areas. The design of a car that produces zero

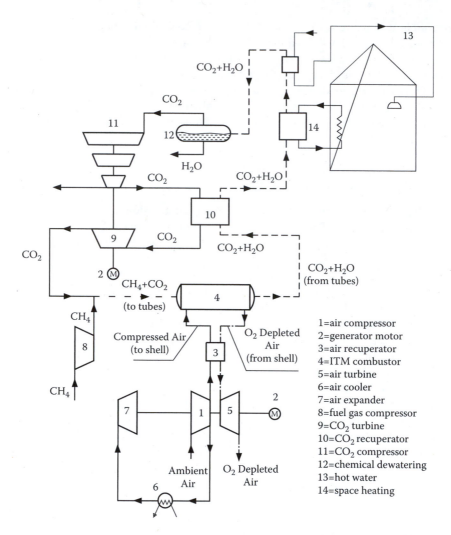

FIGURE 5.4 ZEITMOP cycle for trigeneration (power, heat and cooling) (Yantovsky, unpublished works).

emissions is the most important and most difficult project in relation to efforts to protect the atmosphere. The main problems associated with hydrogen-fueled cars are a low energy volume of hydrogen and an increase in the NO_x formation. The mass and volume restrictions of a car engine are also obstacles in the way of using an ITM combustor.

The first demonstration units therefore for a zero emissions transport power unit should be a ship or a bus engine as the highest possible O_2 flux is needed for such engines. Good efficiency without the need for large recuperators is possible if high

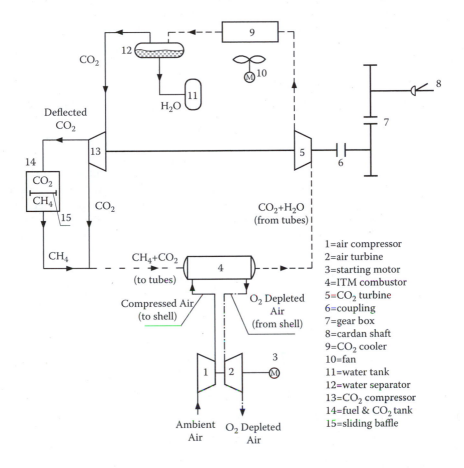

FIGURE 5.5 Application of ZEITMOP for transport (Yanovsky, 2006, unpublished works).

pressures (close to 100 bars) can be sustained on the shell and tube sides of the ITM combustor. An engine outline is shown in Figure 5.5.

Ambient air enters the adiabatic compressor 1, before entering the shell side of the ITM combustor 4. After being heated by the flue gases on the tube side of 4, the oxygen-depleted air expands in turbine 2 before being discharged through the car exhaust as a harmless effluent. The combustion products from the tube side (i.e., permeate side) of 4 are expanded in turbine 5, then cooled by ambient air in the radiator 9, de-watered in 12, and finally, the remaining gaseous CO_2 is compressed in 13 up to about 100 bars. Pressurized liquid CO_2 should be stored in a tank to be discharged at a gas filling station when new fuel is pumped into the tank 15. The sliding baffle is used to keep the CO_2 and fuel in separate compartments. The use of compressed methane as a fuel allows the easiest implementation of this type of scheme. There are many millions of cars in the world today powered by such a fuel. The CO_2 from the tank 15 might be sequestered or, better, used for the enhancement of oil or gas recovery.

5.5 A ZERO EMISSIONS AIRCRAFT ENGINE

In the not-too-distant future, the concept of an ITM combustor could also be applied to produce a zero emissions aircraft. The emissions from aircraft seem negligible when compared to the emissions produced by cars. In the future, however, emissions from aircraft might be found to pose problems for the high layers of the troposphere.

In relation to aircraft, the main drawback with the ITM combustor solution is the much heavier mass of carbon dioxide (compared to that of fuel) that has to be stored and carried by the aircraft, thus reducing the possible payload. The mass of fuel on a commercial jet aircraft is typically about 30% of the total aircraft weight. As the mass of CO_2 will be about three times heavier than the fuel, it would be impossible to carry on board such a mass of combustion gases. Perhaps, detachable containers might be used, at least for a transatlantic flight, enabling them to be jettisoned from the aircraft when flying over the ocean, descending to the ocean floor without any harm to nature.

The aircraft gas turbine engine shown in Figure 5.6 has an ITM combustor embedded within its ordinary aircraft combustor. The heat released from the combustion

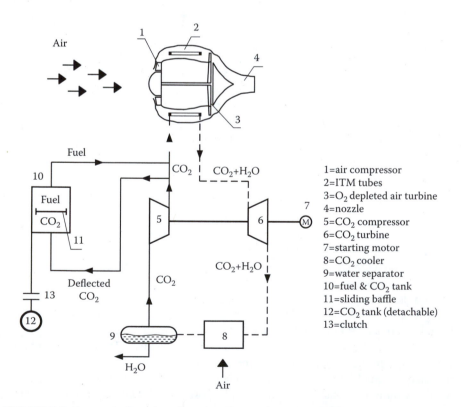

FIGURE 5.6 Zero emissions aircraft engine (Yantovsky, unpublished works).

reaction taking place inside the tubes heats the compressed air through the walls of the tubes. This oxygen-depleted air in the reactive jet stream forms the only emission from the aircraft. It is completely harmless to the atmosphere. The liquid CO_2 is stored in tank 10 and is jettisoned with container 12 on approximately 3 to 5 occasions during a flight across the Atlantic.

5.6 A MEMBRANE SMOKELESS HEATER

Innumerable small fuel-fired heating devices exist, which emit harmful gases into the environment. When converted to a zero emissions cycle, such units do not require a very high pressure in the cycle. CO_2 is required at high pressure only to allow its liquifaction. As this case has fewer engineering problems than those mentioned so far, it may be used for the first practical demonstration of the ITM combustor. This scheme is shown in Figure 5.7. It differs from Figure 5.5 in that it does not have any turbine and all the fans, as well as the CO_2 compressor, though small, are driven by electric motors. The pressure in the ITM combustor, 4, is approximately 1 bar and the temperature is 800 to 900°C. Such conditions have been tested extensively by ten Elshof (1996) and Ritchie et al. (2004), with the $La_{0.5}Sr_{0.5}Fe_{0.8}Ga_{0.2}O_{3-\delta}$ membranes

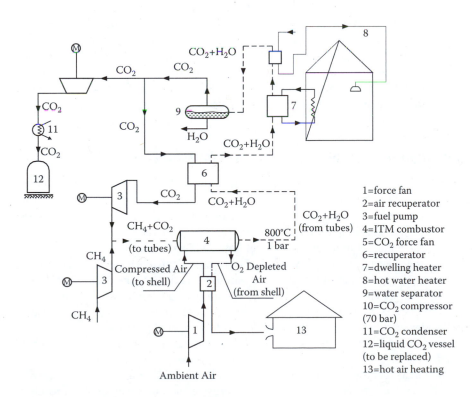

FIGURE 5.7 Smokeless heating system (Yantovsky and Gorski, 2007, unpublished works).

spray deposited on porous alumina tubes. The scheme is quite clear, as shown in Figure 5.7. Liquid CO_2 is stored and collected regularly by the municipal authorities to be disposed of as municipal waste.

As the unit produces rather hot air, it might be used for open space heating (e.g., a hall), mixing the air with the ambient temperature, if required.

5.7   A ZERO EMISSIONS RANKINE CYCLE

In previous sections, the ITM combustor was under high temperature and pressure. This poses significant engineering problems associated with ITM stability and shell strength. Of course, high temperatures and pressures are justified in trying to achieve high power unit efficiencies.

If we allow the efficiency to be somewhat compromised, we can operate the ITM unit under more relaxed conditions, similar to experiments conducted by ten Elshof (1996) and others. The proposed scheme is outlined in Figure 2.14.

It is comprised of an ordinary Rankine cycle 4-5-6-7 where steam is raised in a boiler 4 by the flow of permeate side flue gases from the ITM combustor 3 at ambient pressure and a temperature of about 900°C. After 4, the flue gas mixture, $CO_2 + H_2O$, is cooled in 2 and 8, de-watered in 9, and returned to the ITM combustor via a recuperator 2. The pressure in the CO_2 loop is almost constant with the selection of pressure depending upon the particular case under consideration. A fan for driving CO_2 around the circuit is not indicated in the scheme. Note that as 1 is a fan and not a compressor, the temperature rise of the air through 1 will be negligible. The O_2-depleted air (D-air) leaving 2, therefore, should be quite cool (50–100°C) so that any heat loss with this exit stream should be minimal. A fraction of the CO_2 is deflected, liquified, and sequestered or used elsewhere.

5.8   BOILER INTEGRATED WITH ITM COMBUSTOR

Here, Figure 5.8 is modified to integrate the boiler and the ITM combustor. In contrast to the ITM combustor shown in Figure 5.2, which is similar to an ordinary shell-and-tube heat exchanger, the scheme shown in Figure 5.8 proposes to combine the boiler and ITM combustor by embedding the boiler tubes inside the permeate side tubes while accommodating the fuel flow and oxidation.

For the sake of clarity, only one boiler tube is depicted inside one ITM tube, all encased by one shell tube. It forms only one module of the Membrane Boiler (Memboiler for short). In practice, such a system should contain many modules. The cross-section of the shell of the module might not be circular but square or hexagonal as depicted in Figure 5.8. A bundle of such hexagonal tubes with the boiler and membrane tubes inside forms a honeycomb structure, similar to the honeycomb of wind tunnels.

When the fuel gas, CH_4, is oxidized as a result of O_2 diffusing through the ceramic membranes, heat energy is released. This energy release, however, does not result in

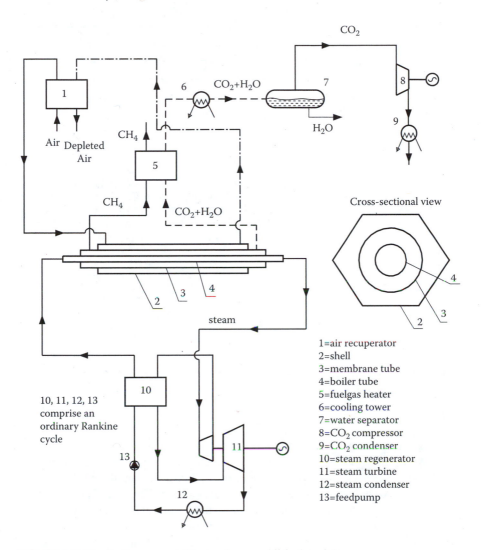

1=air recuperator
2=shell
3=membrane tube
4=boiler tube
5=fuelgas heater
6=cooling tower
7=water separator
8=CO_2 compressor
9=CO_2 condenser
10=steam regenerator
11=steam turbine
12=steam condenser
13=feedpump

10, 11, 12, 13 comprise an ordinary Rankine cycle

FIGURE 5.8 Memboiler concept (Yantovsky, unpublished works).

an increase in the flue gas temperature. Instead, the heat is immediately absorbed through the walls of the boiler tubes by the expanding steam. The tube, in fact, works as a once-through boiler. As the rate of heat transfer from the flue gases on the permeate side will greatly exceed the rate of heat transfer to the shell side, some ribbons or baffles should be installed. The baffles might be helical in shape to increase the path length of the flow of fuel and permeate gases. Increasing the path length in this fashion will help to complete the combustion of the fuel. Helical baffles might also be used on the shell side. The baffles should be made of stainless steel and be flexible enough (e.g., of semi-circular or "C" cross-section) to withstand any heat-induced deformation of the ceramic.

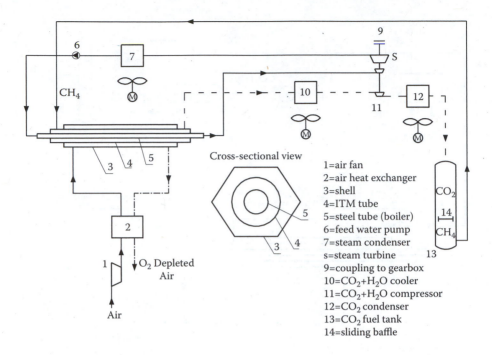

CH$_4$

Cross-sectional view

1=air fan
2=air heat exchanger
3=shell
4=ITM tube
5=steel tube (boiler)
6=feed water pump
7=steam condenser
s=steam turbine
9=coupling to gearbox
10=CO$_2$+H$_2$O cooler
11=CO$_2$+H$_2$O compressor
12=CO$_2$ condenser
13=CO$_2$ fuel tank
14=sliding baffle

O$_2$ Depleted Air

Air

CO$_2$

CH$_4$

FIGURE 5.9 AMSTWEG cycle (automotive membrane steam turbine without exhaust gases).

A slightly different system for a Memboiler has been patented by Praxair. In this model, the steam tubes are located not inside the membrane ceramic tubes, but in a series, (interspersed) with the flame flow directed across the bundle of tubes.

The air feed system does not require a compressor. The hot O$_2$-depleted air exhausted from the system might be used elsewhere. Carbon dioxide (the total flow) is de-watered, liquified, and sequestered or used elsewhere. A version of automotive cycle AMSTWEG (see Figure 2.24) with ITM tubes in the memboiler is presented in Figure 5.9.

REFERENCES

Allam, R. et al. 2000. "Air separation units, design and future developments," in *Proc. ECOS'2000*, July 5–7, Enschede, the Netherlands, P. 4, 1877–1888.

Förster, M., Hannes, K., and R. Telöken. 2001. Combined cycle power plant with pressurized pulverized coal combustion. *VGB Power Technology* No. 9: 30–35.

Korobitsyn, M. et al., 2000. "SOFC as a gas separator," *ECN-Fuels, Conversion & Environment*, Report ECN-C-00-122.

Levin, L., Nesterovski, I., and E. Yantovsky. 2003. "Zero emission Rankine cycle with separate ion transport membrane combustor," presented at *7th Int. Conf. for a Clean Environment*, July 7–10, Lisbon, Portugal.

Ritchie, J.T., Richardson, J.T., and D. Luss. 2004. Ceramic membrane reactor for synthesis gas production, *AIChE J*. 47(9): 2092–2101.

ten Elshof, J.E. 1997. *Dense Inorganic Membranes*. Ph.D. thesis, University of Twente, Enschede, the Netherlands.

Yantovsky, E., J. Gorski, B. Smyth, and J.E. ten Elshof. 2003a. "Zero Emissions Fuel-Fired Power Plant with Ion Transport Membrane," presented at *2nd Annual Conf. for Carbon Sequestration*, May 5–8, Alexandria, VA, U.S.A.

6 Detailed Simulation of the ZEITMOP Cycle

6.1 TURBOMACHINERY FOR CARBON DIOXIDE AS A WORKING SUBSTANCE

The quasi-combined cycle on CO_2, calculated in Chapter 3, was supplied with oxygen from an external source, not integrated within the unit schematic. The bottoming air cycle here is absent. In this chapter, more rigorous calculations should be undertaken as the subject, the ZEITMOP cycle, contains an ion transport membrane (ITM) oxygen reactor and bottom cycle. The working substance here is not pure CO_2 but a mixture with a percentage of steam before entering the turbine, which affects the thermodynamic properties. As these conditions produce rather high temperatures, the working substance is considered a mixture of ideal gases. The process in the compressor, however, is complicated and all the thermodynamic equations of Section 3.1 should be taken into account.

Gas turbines, compressors, and combustors, applicable for zero emissions power plants, require the best flow development achieved up to now in gas turbine technology. Compared to air, the gas has significantly lower values of the gas constant, velocity of sound (~270 m/s), and the ratio of specific heats. To achieve the dynamic similarity with an air compressor, the CO_2 compressor should run with approximately a 25% reduction in the blade tip speed (at the same tip Mach number) and a 15% increase in the mass flow. Therefore, in comparison with other turbomachinery, the CO_2 compressors and turbines are very compact and efficient.

It is obvious that highly effective turbomachinery can substantially improve the cycle potential. In this section, we will deal with recent data on turbomachinery that operate by utilizing CO_2 in order to acquire firm data on the attainable isentropic or polytropic efficiencies.

Generally, for large power applications, the axial-flow turbomachinery (compressors and turbines) are selected. The main reason for employing axial multistage machines is their relatively low volume capacity and less efficiency of the centrifugal machines. Another important aspect of turbomachinery design is synchronization with the grid and setting of the rotational speed.

In Chapter 2, the work of Keller and Strub (1968) was mentioned on the experience of Escher Wyss Co. in the CO_2 machinery construction. It was a successful activity for nuclear energy conversion. Even before, in 1940, D.P. Hochstein proposed CO_2 as a substitute for water in the Rankine cycle with an ordinary boiler on organic fuel, which led to the Hochstein cycle attracting the attention of the nuclear power industry.

For a project of 300 MW_e, the power unit in Gong et al. (2006) and the data concerning the compressors is as follows.

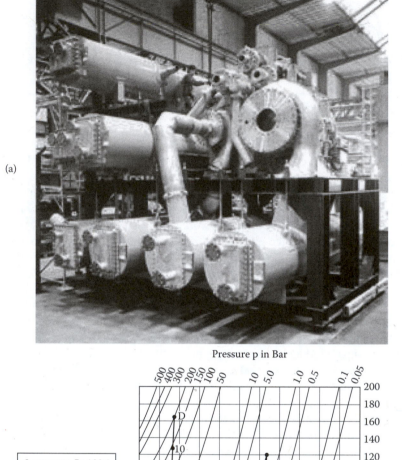

FIGURE 6.1 (a) MAN Turbo 10-stage, 200:1 CO_2 compressor, (b) MAN compressor RG053/10 and its T-s diagram (10-stages, 200:1, CO_2) (www. manturbo.com/en/700/).

Axial-main compressor at 32°C: mass flow rate 2100 kg/s, fluid density 650 kg/m^3, rotational speed 3600 rpm, diameter 80 cm, number of stages 7 (for 32°C at the inlet), discharge pressure 20 MPa, pressure ratio 2.6, and the efficiency 89%. By changing the isentropic efficiency of the compressor from 0.70 to 0.85 and the turbine to 0.90, the cycle efficiency increases from 43.5 to 46.5%.

This is the data of a centrifugal compressor already tested by Mitsubishi Heavy Industries (Sato et al., 2004): type radial, 7 stages, inlet/outlet pressure 0.1/20.3 MPa, mean pressure ratio 2.14, suction volume flow 10.3 m^3/s or mass flow 21.6 kg/s, power input 11.7 MW, discharge gas density 291 kg/m^3 by 172.7°C, efficiency 85%.

The powerful, integrally geared CO_2 compressor manufactured by MAN (see Figure 6.1) displays a clear T-s diagram showing its work in the supercritical region, not transgressing the saturation line, with intercooling after the first seven stages but with none for the last three. All other known industrial CO_2 compressors contain about ten stages.

Recently, a new concept has appeared based on the progress in the design of transonic compressor stages for the aeronautical applications. A supersonic compressor can provide a radical reduction of the stage numbers and intercoolers, together with mass and cost, and such a system has been the dream of aviation and energy engineers since the first flight of the jet aircraft. Now the dream seems to be a reality due to the success of Ramgen Power Systems (Lawlor and Baldwin, 2005).

A diagram of the Rampressor™ is shown at Figure 6.2 and below is a schematic of the shock wave which is similar to the entrance of an inlet of a supersonic aircraft engine and the Rampressor.

This similarity inspired Arthur Kantrovitz, the head of AVCO Laboratory, to propose "*the supersonic axial-flow compressor*" as NACA Report ARC No. L6D02 in 1946. The main idea of Ramgen is the rotor configuration, which is without blades and a rotating system of oblique shocks ended by rather weak normal shock. In such a system, the total pressure is almost recovered due to the smooth deceleration of the compressible flow. The Rampressor has been tested on air to 21,300 rpm with an inlet of about Mach number two. The test results confirmed the calculations to a high degree of accuracy. Basic performance parameters of the Ramgen compressor and conventional centrifugal machines have been compared by Lawlor et al. (2005); see Table 6.1.

It can be seen that the input power requirement for a Ramgen compressor is very reasonable and offers the additional net effect of heat recovering of gas compression (over 70%), significantly improving the energy economy. Such compressors would be less expensive, more compact, and more efficient than competing conventional designs.

On the basis of previous work on the Graz cycle development, Jericha et al. (2006) presented the design concept for turbomachinery components of 400 MW$_e$ net power output. This power is derived from a 490 MW turbo shaft unit and the design concept for this size is presented with a two-shaft configuration. A fast running compression shaft is driven by the high-pressure compressor turbine (HPCT), whereas the power

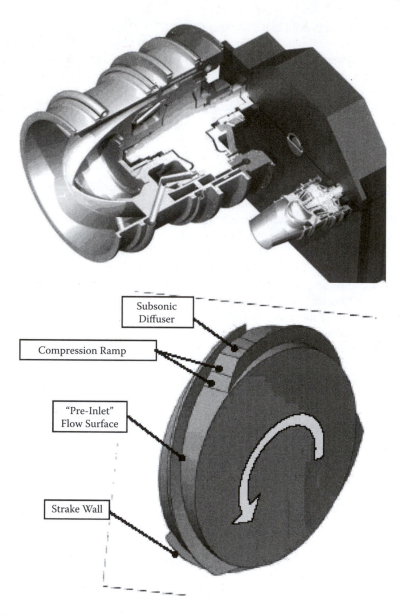

FIGURE 6.2 Rampressor™—A supersonic CO_2 compressor and its shock waves system (Gong et al., 2006).

shaft is comprised of the power turbine (HPT) and the low pressure steam turbine (LPST).

The CO_2 turbines seem to be proven following Escher-Wyss tests and do not pose any serious problems but for the ability of materials to withstand high temperatures.

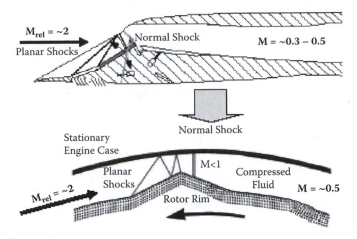

FIGURE 6.2 (Continued).

In general, the conclusion is that the equipment for ZEITMOP cycle (compressors, turbines, ITM oxygen reactors) is available, at least for a demonstration plant. Figures of isentropic efficiencies for cycle calculations, similar to those assumed in the following sections, are quite correct.

TABLE 6.1

Comparison of Ramgen Compressor to Conventional Turbomachinery

	Compressor Type		
Parameter	**Ramgen**	**IGRC[a]**	**ILPC[b]**
Mass flow, kg/h	68,000	68,000	68,000
Inlet volume flow, m³/h	36,360	36,360	36,360
Number of stages	2	8	12
Intercoolers	1	7	2
Casings	1	1	3
Power, MW	7,333	7,382	8,312
Isothermal efficiency	65.8%	64%	56.9%
Stage/casing discharge temperature, °C	243	99	193
Max. thermal recovery temperature, °C	120	120	120
BHP/100	45.9	46.2	52.1
% recoverable	71.8%	7.5%	50.2%
Net kW	2.070	6.828	4.141

[a] IGRC–Integrally geared radial compressor
[b] ILPC–In-line process compressor

Source: Lawlor, S., 2007.

6.2 ZEITMOP CYCLE ANALYSIS

In the data previously presented, an extensive analysis of the ZEITMOP cycle was examined based on an assumption that the constants limiting pressure in the cycle were close to 22 MPa. Such a high pressure level, as well as the maximum temperature range, was taken from the conclusions formulated from early data assumed in the COOPERATE and MATIANT cycles which are forerunners of the ZEITMOP concept cycle; see Figure 6.3.

New and more consistent results have been obtained within a standard Aspen Plus® v.11.1 simulation environment supported by additional procedures for the analysis of atypical components and chemical processes in the system. The steady-state simulation of the cycle was connected directly to a simplified model of a mixed-conducting ion transport membrane for the separation of oxygen from the atmosphere (ITM unit). This analysis is based on ten Elshof's (1997) data for the performance of $La_{1-x}Sr_xFeO_{3-\delta}$ dense membranes and the Excel® worksheet calculations.

Some simulation tests were performed in order to verify the Aspen Plus library for the calculation of fluid properties in the cycle. This important element of the calculations was conducted in order to select a convenient EOS (equation of state) describing both pure CO_2 properties and their combination with steam or oxygen, circulating in the cycle. The EOS given by Peng-Robinson-Boston-Mathias (PRBM) showed their flexibility and good quality results in the widespread range of state parameters and gas compositions (including a liquid phase transition). It delivered more valuable information on the performance cycle and each unit operation process compared to the early steps (Yantovsky et al., 2004) and formulae developed and calculated manually aimed at a CO_2 standard and the modified virial EOSs.

The present version introduces a heat exchanger HE3 in the main pipeline where the pure CO_2, after expanding in the HPT turbine (T-CO2), is heated by the depleted hot air stream leaving the ITM separator (Figure 6.3). A high temperature mixture (~700°C) of

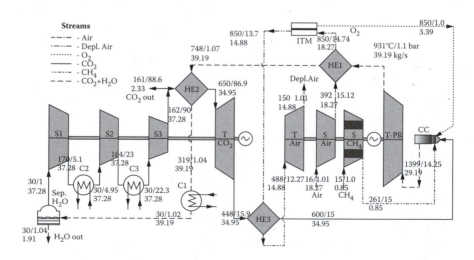

FIGURE 6.3 The modified ZEITMOP schematics (Yantovsky et al., 2006).

CO_2 and oxygen is delivered to the combustion chamber (CC). It is easy to see that this modification provides a better reheating of the CO_2 recirculating stream and reduces fuel consumption in CC for an assumed range of combustion gases temperature. The positive effect of heat recovery from the depleted air stream is a good way to obtain better control of the energy utilization in a main cycle and avoid excessive loss caused by waste energy flow from the depleted air to an ambient surrounding.

The simulation of the ZEITMOP cycle has shown the important influence that limiting the range of the state parameters has on performance and efficiency. Starting from the assumed version of an idealized case when the pressure and heat losses were neglected, the simulation of the cycle has delivered some background for a proper selection of pressures, temperatures, and fluid streams. It was found at this stage of the analysis that the primary rule in cycle performance factors should be expected from a broad range variation of the ultimate pressures and temperatures. The cases analyzed were carried out within a range of cycle temperatures between 1200 to 1500°C and pressure from 8 to 22 MPa.

At these top cycle conditions, the bottom limit of thermal parameters was taken close to the atmospheric boundaries (0.1 MPa, 30°C). Several cycle simulation tests between the limiting state parameters mentioned have shown an almost linear dependence of the net power production and thermal cycle efficiency with respect to the upper limit of the temperatures. For the upper range of pressure variation, an opposite tendency was noticed up to the assumed reasonable data for an operation of the energy conversion units (pressure ratios for turbine and compressor stages, energy density streams in the heat exchangers and reactors, ITM operating limits, etc.). As in the analysis previously conducted (Yantovsky et al., 2004), the calculated system effectiveness is based on an actual and typical data set for compression and expansion processes in modern turbomachinery.

The subsequent step of using steady-state process simulation oriented on the 1-th Law principles was dealt with based on typical data on the fluid-flow behavior of modern regenerative heat exchangers and chemical reactors. The pressure losses equal to 1–3% of the inlet pressure for the particular unit were assumed. The high-temperature mixing losses caused by the cooling flows in high temperature turbine blade passages gave an additional and very strong effect on the cycle performance.

According to recent work on aero-gas turbine cooling (Mattingly et al., 2002), these losses change the turbine efficiency η_{t0} approximately proportional the relative coolant flow ε_{ct} and gas temperature T_g [K]

$$\Delta \eta_t \cong \eta_{t0} - 0,9 \cdot \varepsilon_{ct},\tag{6.1}$$

where

$$\varepsilon_{ct} = m_{ct}/m_g \cong (T_t - 1330)/8800 \tag{6.2}$$

In this approach, only typical conditions for the gas cooling system have been taken (i.e., cold CO_2 stream from the compressor S2 bleed). New advanced and more sophisticated cooling turbine systems have recently been developed (steam cooling), but they were not considered in this work. It should be noticed that the resulting reduction of cycle efficiency is simply proportional to these cooling losses.

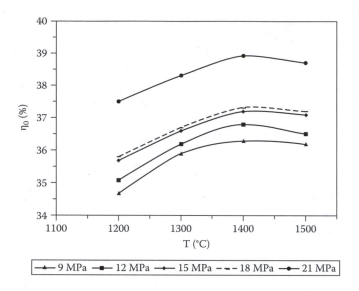

FIGURE 6.4 Cycle efficiency versus temperature (without heat exchanger HE3).

After introducing the sequential heat regeneration of the mean pressure CO_2 stream (regressing to the ITM and combustion chamber) heated by the depleted air behind the ITM, a reasonable improvement of cycle efficiency will be obtained (up by 5 to 7 percentage points).

This is caused by the smaller energy needs for a chemical reaction and the heating of combustion products in the combustion chamber (a lesser amount of fuel necessary as well as oxygen and the primary air). The reduced quantity of air creates a lower energy demand for an air compressor (but subsequently, less effective expansion power in an air turbine). In the total energy balance of the cycle, one can observe a growth in the cycle efficiency.

These results for the maximum cycle pressures of 9 to 21 MPa and limiting upper temperatures from 1200 to 1500°C analyzed are presented in Figure 6.4 and Figure 6.5.

Such data are compared for two basic cases—the first without the regenerator HE3 (as was discussed by Yantovsky et al., 2004), and the second including the regenerative heat exchanger. The thermal efficiency growth observed (based on LHV for gas fuel methane) and their extreme existence meet the optimum operational conditions for the high temperature turbine T-PR (Figure 6.4) and a "penalty" effect caused by the turbine cooling requirements (see Figure 3.3 and Figure 3.4). Good performance conditions for the ZEITMOP cycle will be expected very close to maximum pressure $p_{max} = 9$ MPa and TIT = 1400 to 1430°C.

It should be mentioned that the actual data are less optimistic than the results obtained previously. The present case was simulated with similar assumptions as before and the selected results obtained of the final calculation are presented in Table 6.2 and Table 6.3.

A high heat energy output level and net turbine power in all the conditions analyzed was found but the attached data in Table 6.2 and Table 6.3 only refer to the low bound of

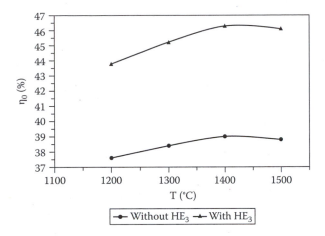

FIGURE 6.5 The increase of cycle efficiency due to the inclusion of HE3.

principal state parameters in the cycle (1200°C and 9 MPa). It is easy to see that the introduction of the heat exchanger HE3 into the recirculated CO_2 loop in this case (1200°C and 9 MPa) gives a favorable temperature growth of the incoming $CO_2 + O_2$ stream close to 167°C and a subsequent energy saving in the combustor CC. The resulting cycle efficiency (see Figure 6.5) will be approximately 5.8% higher than presented in Figure 6.4 and the fuel consumption is about 18% lower. This relation poses a very practical significance for further tests and more detailed analysis. It is a starting point to the introduction of exergy analysis of all the system components as well as the complete cycle.

An important conclusion should be formulated based on these ultimate results—the cycle thermal efficiency obtained has a reproducible tendency (Figure 6.5), but

TABLE 6.2
ZEITMOP Turbine and Compressor Loads

Block	Break Power [kW]	Mass Flow [kg/s]	Efficiency	Isentropic Head [kJ/kg]	P_{in} [bar]	P_{out} [bar]
T-AIR	6,124.14	17.818	0.890	386.19	12.73	0.89
T-PR	21,394.72	38.143	0.895	626.20	14.25	1.10
S-CH$_4$	404.20	0.640	0.870	549.45	1.00	15.00
T-CO$_2$	7,664.49	34.949	0.880	249.21	85.90	15.00
S1	4,682.06	36.707	0.870	110.97	1.00	5.10
S2	4,278.84	36.707	0.870	101.41	4.95	23.00
S3	3,563.87	36.707	0.850	82.53	22.30	90.00
S-AIR	8,087.53	20.371	0.850	337.46	1.01	15.01

Source: Yantovsky, E. et al., 2004.

TABLE 6.3
ZEITMOP Streams Data (1200°C/9 MPa)

Stream	Flow [kg/s]	Temperature [°C]	Pressure [bar]	Enthalpy [kJ/kg]	Entropy [kJ/kgK]
AIR-1	20.37	15.0	1.013	10.44	0.1041
AIR-2	20.37	396.7	15.01	396.57	0.1969
AIR-3	20.37	760.0	14.74	793.58	0.6864
AIR-4	17.82	760.0	12.73	800.22	0.6865
AIR-4A	17.82	424.9	12.73	421.11	0.2830
AIR-5	17.82	100.3	1.010	77.39	0.3540
B-TLEN	2.554	760.0	0.130	745.50	0.6703
C-MIX	37.50	577.0	13.0	774.22	0.6880
CH_4-1	0.640	15.0	1.0	36.69	0.4938
CH_4-2	0.640	260.8	15.0	668.24	05096
E-PR1	38.14	1199.5	14.25	1,152.26	1.6324
E-PR2	38.14	780.5	1.10	594.52	1.6888
E-PR3	38.14	608.2	1.07	548.59	1.4692
E-PR4	38.14	200.4	1.04	168.82	1.5408
E-PR5	38.14	30.0	1.01	31.48	1.6395
H_2O	1.346	30.0	1.01	125.69	0.4366
CO_2 + 1	36.71	30.0	1.00	29.37	0.0827
CO_2 + 2	36.71	170.1	5.10	156.92	0.1209
CO_2 + 3	36.71	30.0	4.95	29.27	0.0927
CO_2 + 4	36.71	164.4	23.0	145.83	0.1279
CO_2 + 5	36.71	30.0	22.3	29.79	0.0719
CO_2 + 6	36.71	161.6	90.0	125.88	0.1130
CO_2 + 7	34.95	161.0	88.6	125.89	0.1131
CO_2 + 8	34.95	590.0	85.9	568.62	0.4034
CO_2 + 9	34.95	397.5	15.0	346.78	0.4489
CO_2 + 9A	34.95	565.0	15.0	531.23	0.5573
CO_2 + OUT	1.758	161.1	88.6	125.88	0.1130

is 5 to 8% higher and the fuel consumption is reduced by up to 20% in comparison to the basic case without the regenerator HE3. Other than this, the final state parameters after depleted air expansion in a turbine S-AIR should be closer to the normal atmospheric conditions. From a technical point of view, it is possible to arrange the ZEITMOP cycle in a pilot scale plant. The progress in the micro-cogeneration systems is sufficient for developing a demonstration power unit of 100 to 200 kW.

6.3 ZEITMOP CYCLE WITH COMBINED COMBUSTION CHAMBER AND ITM REACTOR

The original cycle (Yantovsky et al., 2004), included an oxygen transport membrane (OTM) air separation unit and a separate combustion chamber. The results of a detailed Aspen Plus v.10 simulation of this cycle, at combustion temperatures between 1200 and 1500°C, with the OTM unit operating at temperatures of 750 to 1000°C, were presented at the ECOS 2006 conference by Yantovsky et al. This layout will be referred to here as ZEITMOP-separate.

A different possible layout for the cycle is to combine the combustion chamber with the air separation unit. The results of initial calculations based on a simplified model of the combined cycle for combustion and OTM operation at temperatures of 1000 and 1400°C were also presented at the ECOS 2006 conference by Foy and McGovern.

This layout will be referred to as ZEITMOP-combined. This section presents results of a detailed Aspen Plus simulation of the ZEITMOP-combined option operating at temperatures between 900 and 1500°C (see Foy et al., 2007).

The temperature of the OTM would be equal to the combustion temperature, which is higher than the temperature that could otherwise be achieved in the OTM. The oxygen would be consumed quickly after traveling through the membrane, maintaining a low oxygen partial pressure on the combustion side of the membrane.

Both of these factors would increase the oxygen flux per unit of membrane area. The combined unit would therefore be much smaller physically than the air separation unit in the original layout. The combustion chamber would also be eliminated.

This reduction of the balance of the plant offers a clear motive for investigating the effect of this new layout on the cycle.

The temperature at which combustion occurs will be determined by the limits of the OTM material. Industrial research has recently begun into developing OTM materials specifically for combustion (see van Hassel et al., 2005). At present, the maximum temperature these materials can withstand is 1000°C; however, this is likely to increase.

Foy and McGovern (2006) concluded that at high combustion temperatures, the efficiency of the ZEITMOP-combined cycle is similar to the efficiency of the ZEITMOP-separate cycle, but that the combined option at low combustion temperatures has a lower efficiency than the efficiency of either option at higher combustion temperatures.

In other words, the combined option was found to be worthwhile only if the OTM/combustion unit can operate at high temperatures. Its schematics are given in Figure 6.6.

Methane (D-CH$_4$-1) is compressed (D-CH$_4$-2) and enters the OTM unit, where it is burned with oxygen in the presence of carbon dioxide. The combustion products (E-PR-1) are expanded (E-PR-2) and cooled in two heat exchangers (E-PR-4) and a cooling tower, which reduces the temperature to 30°C, causing the water to condense (E-PR-5). The liquid water is then separated and removed (F-H$_2$O), leaving pure carbon dioxide (G-CO$_2$-1) which is compressed with intercooling (G-CO$_2$-6).

The portion born of combustion is removed (G-CO$_2$-OUT), while the rest is heated and expanded before returning it to the OTM unit (G-CO$_2$-9). Air (AIR-1) is compressed and heated before entering the OTM unit (AIR-3). The hot, oxygen-depleted air is expanded and cooled before re-entering the atmosphere (AIR-7).

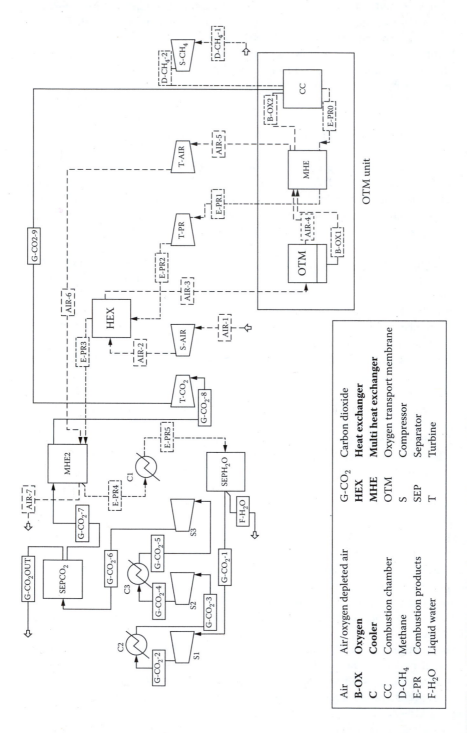

FIGURE 6.6 ZEITMOP-combined cycle model (Foy et al., 2007).

6.4 SIMULATION OF OXYGEN TRANSPORT MEMBRANE UNITS

Oxygen transport membranes (OTMs) are ceramic membranes that contain oxygen ion vacancies in the ceramic lattice (see Chapter 4). When heated, excited oxygen ions can travel through the lattice using these vacancies.

There are two main types of ceramics with oxygen ion transport capabilities: perovskites and fluorites. Perovskite oxygen transport ceramics have the chemical structure ABO_3. Fluorite oxygen transport ceramics have the structure AO_2 (e.g., ZrO_2).

The oxygen flux across an OTM is given by the Nernst-Einstein equation, which shows that the flux rises with the difference in oxygen partial pressures across the membrane, and can be particularly high when an oxygen-consuming reaction occurs on one side of the membrane.

The flux is also proportional to the temperature, so an exothermic oxygen-consuming reaction should provide a very high oxygen flux. Combustion seems ideal. However, ceramic sinters and eventually melts at high temperatures, so every OTM material has a maximum operating temperature. Perovskite materials typically have maximum temperatures of about 800 to 1000°C. Fluorite materials can withstand far higher temperatures; however, fluorite materials have lower oxygen fluxes than perovskites.

Development of new OTM materials is ongoing by a number of companies and groups around the world. It is probable that OTM materials will be developed with a maximum temperature of above 1400°C, and a relatively high flux. Modified fluorite membranes may be the key to this development.

The OTM unit in the ZEITMOP cycle will most likely consist of a bank of tubes, much like a shell and tube heat exchanger (see Figure 5.2), with air on one side and carbon dioxide on the other. Warchol (in Foy et al., 2007) modeled the OTM as a unit which removed some oxygen from the air, operating at temperatures between 750 and 1000°C.

The oxygen stream is then joined with a CO_2 stream. This means that the situation shown in Figure 6.7 (data for combustion temperature 1400°C) was modeled as shown in Figure 6.8. This model therefore assumed that the temperature of the OTM material was 925°C, whereas the actual average temperature of the OTM material is closer to 780°C.

The reduction in flux will only affect the physical size of the OTM unit and not the efficiency of the cycle, so this does not affect the validity of the model. However, this simplified model also does not allow for any heat transfer in the OTM unit itself. While this is acceptable for a low temperature ratio, it is an unacceptable assumption

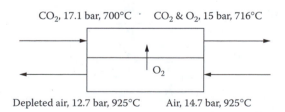

CO₂, 17.1 bar, 700°C   CO₂ & O₂, 15 bar, 716°C

O₂

Depleted air, 12.7 bar, 925°C   Air, 14.7 bar, 925°C

FIGURE 6.7 OTM unit inlets and outlets in ZEITMOP-separate (Foy et al., 2007).

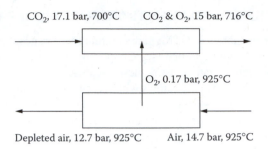

FIGURE 6.8 Simplified OTM model used for ZEITMOP-separate (Foy et al., 2007).

when the combustion occurs on the permeate side of the OTM, i.e., for ZEITMOP-combined, as the unit is likely to operate as a heat exchanger.

In fact, initial calculations based on the thermal conductivity, thickness, and area of the OTM material assumed, show that at the temperatures investigated, the unit will operate as an almost perfect heat exchanger. In the simulations presented in this paper, the OTM/combustion unit is modeled as a unit which separates O_2 from air, a heat exchanger, and a combustion chamber. The situation shown in Figure 6.9 is therefore modeled as shown in Figure 6.5.

The results presented by Yantovsky et al. (2006) were for combustion temperatures between 1200 and 1500°C, and delivery pressures for carbon dioxide between 90 and 210 bars, with and without the additional heat exchanger.

In this section, we present results of simulations for the cycle configuration in Figure 6.6 only, at the analogous temperature and pressures of carbon dioxide, and also for combustion pressures between 3 and 40 bars when CO_2 is delivered at 210 bars.

The assumptions used in the simulations were:

- The OTM used is based partly on data published by ten Elshof (1997) and partly on data published by Kharton et al. (1999).
- The OTM unit can withstand total pressure ratios in the region of 1.15 across the membrane.

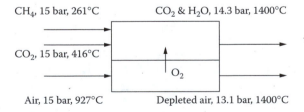

FIGURE 6.9 OTM unit inlets and outlets in ZEITMOP-combined (Foy et al., 2007).

- The OTM unit can withstand temperature ratios in the region of 1.4 across the membrane.
- 98% of CH_4 is converted to CO_2 and H_2O. No other gases appear in the products of combustion.
- Heat exchanger pinch points are 14 to 16 Centigrade.
- The Peng-Robinson-Boston-Mathias equation of state was used to calculate the properties of pure CO_2 and mixtures of CO_2 and either water vapor or oxygen, at all relevant points.
- Pressure losses in heat exchangers and coolers are 3% of the inlet pressure.
- Pressure losses in separators are 1.5% of the inlet pressure.
- Pressure losses in the OTM unit are based on the experimental data (ten Elshof, 1997).
- Compressor isentropic efficiencies are 87%, except for the last stage of CO_2 compression, which has an isentropic efficiency of 85%.
- Turbine efficiency was adjusted for cooling flow as described in Chapter 3. This takes into account the effect of a coolant flow, but without modeling the coolant flow itself. The isentropic efficiencies of the air and combustion product turbines therefore vary from 89.5% at 900°C to 85% at 1500°C.
- Heat is rejected at 30°C.
- The OTM removes 60% of the available O_2 from the air.

6.5 RESULTS AND DISCUSSION

The efficiency of the ZEITMOP-combined cycle for a CO_2 delivery pressure of 210 bars, at various combustion temperatures and pressures, is shown in Figure 6.6. The efficiency varies much more with combustion temperature than with combustion pressure. The efficiency rises with temperature for all combustion pressures. At relatively low combustion pressures (3–10 bars), the efficiency rises smoothly with temperature. At pressure of 40 bars, there is a different shape to the curve, with a sharper increase in efficiency at lower temperatures, and a lower slope at higher temperatures.

At intermediate combustion pressures (15–30 bars), there are two distinct regions in the efficiency and temperature curve. At lower temperatures, the curve resembles the high-pressure curve, while at higher temperatures, the curve resembles the low-pressure curve. By analysis of the temperatures of individual streams, it was found that this is related to the operation of the multistream heat exchanger, MHE2 in Figure 6.10. This transfers heat from both the depleted air stream Air-6 and the combustion products stream E-PR3 to the recirculated CO_2 stream G-CO_2-7. At lower combustion temperatures (below 1000°C for 15 bars, 1100°C for 20 bars, and 1300°C for 30 bars), E-PR3 is hotter than Air-6 at the inlet to MHE2, whereas at higher combustion temperatures, Air-6 is hotter than E-PR3. As the temperature of the CO_2 stream depends on the hotter of the two feed streams, the power output of the CO_2 turbine is directly affected by this change.

A second reason for the change in slopes is that the temperature of combustion directly affects the amount of CO_2 recirculated in the system, as the CO_2 acts as a coolant for the combustion chamber. This difference is more pronounced at lower

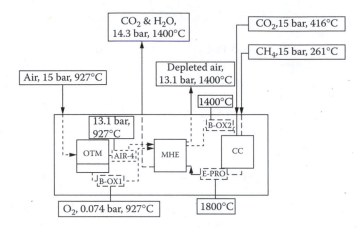

FIGURE 6.10 Simplified OTM model used for ZEITMOP-combined (Foy et al., 2007).

temperatures and pressures in steeper slopes of the curve. Both of these effects directly affect the power produced by the CO_2 turbine, and therefore, the efficiency (see Figure 6.11).

At temperatures of 1150°C and above, the highest efficiency is found at combustion pressures in the region of 3 to 5 bars.

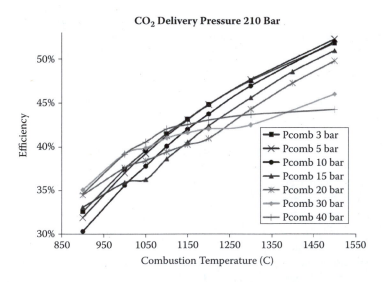

FIGURE 6.11 Efficiency of ZEITMOP-combined when CO_2 is delivered at 210 bars (Foy et al., 2007).

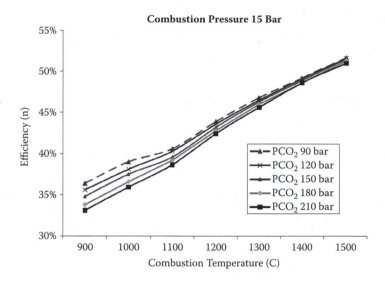

FIGURE 6.12 Efficiencies of ZEITMOP-combined cycle at combustion pressure of 15 bars (Foy et al., 2007).

The efficiency of the ZEITMOP-combined cycle for a 15 bar combustion pressure, at various combustion temperatures and CO_2 delivery pressures, is shown in Figure 6.12.

The efficiency rises as CO_2 delivery pressure drops. This is because less work is required to compress the CO_2.

In a carbon capture and storage situation, the CO_2 would be sequestered and would need to be at a particular pressure depending on the sequestration site. If the cycle delivers liquid CO_2 at a pressure lower than this, it would need to be compressed again.

Calculations presented by Foy and McGovern (2006) imply that this is not worthwhile. It appears that for optimal efficiency, the cycle should be operated at the minimum CO_2 delivery pressure that is suitable for the chosen sequestration option. Compressors for enhanced natural gas recovery which compress CO_2 to 190 bars are available. Therefore, the required CO_2 delivery pressure is likely to be in the range of 180 to 210 bars, depending on a particular gas well or other sequestration option.

The efficiencies of the ZEITMOP-separate cycle with an additional heat exchanger analyzed previously and ZEITMOP-combined cycle are compared in Figure 6.13. At temperatures above 1200°C, the combined ZEITMOP cycle is more efficient than the separate ZEITMOP cycle. Below this temperature, the efficiencies of both cycles are much lower.

The Graz group (Jericha et al., 2004) have developed designs for a turbine that can operate at 1300°C with 77% CO_2, 23% H_2O, so it seems unlikely that the ZEITMOP plant would ever be built with a combustion temperature lower than 1200°C. In this case, OTM materials must be developed that can withstand the combustion temperature.

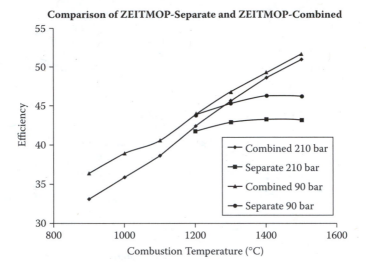

FIGURE 6.13 Efficiencies of ZEITMOP-combined and ZEITMOP-separate cycles (Foy et al., 2007).

If these materials are not available, a high combustion temperature should be maintained by separating the OTM unit from combustion chamber. The results show that:

- The efficiency of ZEITMOP-combined is higher than the efficiency of ZEITMOP-separate at temperatures of 1200°C and above. This improvement increases with combustion temperature from less than 1 percentage point at 1200°C to 5–8% at 1500°C.
- Combining the OTM unit and combustion chamber is worthwhile at combustion temperatures of 1200°C and above.
- If no OTM material can be found to withstand combustion temperatures above 1200°C, a separate combustion chamber and OTM unit will allow the ZEITMOP cycle to achieve a higher combustion temperature and thus, a higher efficiency.
- It has been confirmed that heat exchange from the combustion stream to the carbon dioxide and air streams increases efficiency.
- The optimal conditions for the ZEITMOP-combined cycle are: a combustion temperature as high as possible, combustion pressure 3 to 5 bars, CO_2 delivery pressure as low as possible for the end application.

REFERENCES

Anderson, R.E. et al., 2004. "Demonstration and Commercialization of Zero-Emission Power Plant," in *Proc. 29 Techn. Conf. on Coal Utilization & Fuel Systems*, April 18–22, Clearwater, FL, U.S.A.

Clean Energy Systems, company website, www.cleanenergysystems.com, (accessed Sept. 11, 2007).

Foy, K. and J. McGovern. 2006. "Analysis of the effects of combining air separation with combustion in a zero emissions (ZEITMOP) cycle," in *Proc. of ECOS 2006*, July 12–14 Aghia Pelagia, Crete, Greece, 1693–1701.

Foy, K., Warchol, R., and J. McGovern. 2007. "A Detailed Simulation of the ZEITMOP Cycle with Combined Air Separation and Combustion," in *Proc. of ECOS 2007*, A. Mirandola et al. (eds.), June 25–28, Padova, Italy, 349–356.

Gong, Y. et al., 2006. "Analysis of Radial Compressor Options for Supercritical CO_2 Power Conversion Cycles," *Topical Report, MIT-GRF-034*, Aug. 30, http://nuclear.inl.gov/deliverables/docs/topical_report_mit-gfr-034.pdf (accessed Oct. 19, 2007).

Jericha, H. et al., 2004. Design optimization of the Graz cycle. *ASME Trans., J. Eng Gas Turbines & Power.* 126 (4): 733–740.

Jericha, H., Sanz, W., and E. Göttlich. 2006. "Design Concept for Large Output Graz Cycle Gas Turbines," presented at ASME Turbo Expo, Barcelona, Spain, Paper GT2006-90032.

Keller, C. and R. Strub. 1968. "The gas turbine for nuclear power reactors," presented in *VII World Energy Congress*, Moscow, Paper 167.

Kharton, V. et al., 1999. Perovskite-type oxides for high-temperature oxygen separation membranes. *Jour. of Membrane Sci.* (163): 307–317.

Lawlor, S. and P. Baldwin. 2005. "Conceptual Design of a Supersonic CO_2 compressor," in *ASME, TURBO EXPO 2005*, June 6–9, Reno, Nevada, Paper GT-2005-68349, www.ramgen.com/apps_comp_storage.html (accessed Aug. 15, 2006).

MAN TURBO AG company website, "10-stage integrally-geared compressor for urea synthesis processes," http://www.manturbo.com/en/700/ (accessed Jan. 19, 2007).

Mattingly, J.D. et al., 2002. *Aircraft Engine Design* (2nd ed.). New York: AIAA Edu. Ser.

Mohanty, K.K. 2003. Near-term energy challenge. *AIChE Jour.* 49 (10): 2454–2460.

Renz, U. et al., 2005. "Entwicklung eines CO_2-emissiomsfreien Kohleverbrennungs-processes zur Stromer-zeugung," *RWTH Aachen CCS-Tagung*, Nov. 10–11 Julich, Germany (in German).

Sato, T., Tasaki, A., and J. Masutani. 2004. "Re-injection Compressors for Greenhouse Gas CO_2," Mitsubishi Heavy Industries, Ltd., *Technical Review*, vol. 41, No. 3 (June, 2004).

Sundkvist, S.G. et al., 2001. "AZEP—Development of an integrated air separation membrane—gas turbine," in: *2nd Nordic Minisymposium on Carbon Dioxide Capture and Storage*, Oct. 26, Göteborg, Sweden. www.entek/chalmers.se/~anly/ symp/symp2001 (accessed Nov. 5, 2004).

ten Elshof, J. 1997. Dense inorganic membranes. Ph.D. thesis, University of Twente, Enschede, the Netherlands.

van Hassel, B.A. et al., 2005. "Advanced Oxy-Fuel Boilers for Cost-Effective CO_2 Capture," in *Proc. of 4th Annual Conference on Carbon Capture and Sequestration*, May 5–10, Alexandria, VA, U.S.A., Paper 215.

Warchol, R. 2006. A Perspective on Low Emission Power Cycles Using Carbon Dioxide. Ph.D. thesis, Rzeszow University of Technology, Rzeszow, Poland (in Polish).

Yantovsky, E. et al., 2004. Zero-emission fuel-fired power plants with ion transport membrane. *Energy Jour.* 29 (11-12): 2077–2088.

Yantovsky, E., Gorski, J., Smith, B., and J. ten Elshof. 2004. Zero emission fuel-fired power plant with ion transport membrane. *Energy Jour.* 29 (12-15): 2077–2088.

Yantovsky, E., Gorski, J., and R. Warchol. 2006. "Toward the optimization of the ZEITMOP cycle," in *Proc. of ECOS 2006*, July 12–14, Aghia Pelagia, Crete, Greece, 913–919.

7 Zero Emissions Piston Engines with Oxygen Enrichment

7.1 MAIN CULPRIT

In the section, "Up Front," in *Power Engineering International* (PEi) Jan/Feb 2007 issue, Senior Editor Heather Johnston asks: "Is CCS ready for action?" (CCS = carbon capture and storage). She adds, "People mentioned that the idea of CCS being discussed at a mainstream event such as POWER-GEN International would have been inconceivable as little as one or two years ago." In the next issue of PEi, she indicated, "Hunton Energy, a Texas-based independent power producer, announced its intention to build $2.4 billion IGCC plant that will capture and sequester CO_2." The World Energy Council includes in policy changes "implemented quickly within the electricity sector ... carbon capture and storage (CCS) for power generation ...".

Looking at the modern flood of papers on CCS, we should remember the first proposal by C. Marchetti in 1977 and the documented history of engineering activity (see Chapter 2). We see in the annual conferences in the U.S.A. and international conferences on greenhouse gas control technologies the definitely positive answer to Johnston's question.

However, the fuel-fired power plant is not the only culprit of carbon emissions. Even more dangerous is another one, the world fleet of vehicles. The carbon capture on vehicles is more difficult due to volume and weight restrictions, but the principles of capture onboard are similar to that of a power plant. Concentrating on piston engines, we should try to develop small power plants with cogeneration of heat for district heating to use in densely populated urban areas since these plants have zero emissions.

It is well known that transportation consumes about 70% of U.S. oil and generates a third of the national carbon emissions. Many U.S. and Japanese car manufacturers are actively developing fuel-cell and hydrogen vehicles as the main defense against emissions. Byron McCormick (2003), director of fuel cell activities for General Motors, stated:

> So far, we have reduced the cost of a vehicle fuel cell by a factor of 10. But we have a long way yet to go. We need another 10-times reduction before we can match the cost of today's comparable piston engine.

Assuming that the reduction cost versus time is an exponential decay, the required 10-times reduction may be realized near the end of the century.

Steve Ashley (2005) also stated:

Today between 600 and 800 fuel-cell vehicles are reportedly under trial across the globe. … the motor vehicle industry and national governments have invested tens of billions of dollars during the past 10 years to bring to reality a clean, efficient propulsion technology that is intended to replace the venerable internal combustion engine. … nearly all car company representatives call for more government investment in basic research and hydrogen distribution system … Nobody really knows how to store enough hydrogen fuel in a reasonable volume.

Concerted efforts of numerous companies and labs are continuing at considerable cost in the development of hydrogen cars and fuel cells for clean propulsion systems. But to date, none of the hundreds of experimental cars reveal an ability to store enough energy onboard, comparable to a gasoline or compressed natural gas vehicle. The problem seems to be far from a solution, a concern about which is expressed on the highest political levels.

There exists in Wikipedia (2006) an extensive classification of all the types of vehicles with reduced pollution, reflecting concerted research efforts.

An ultra low emissions vehicle (ULEV) is a vehicle that has been verified by the California Air Resources Board (CARB) to emit 50% less polluting emissions than the average for new cars released in that model year. The ULEV is one of a number of designations given by the CARB to signify the level of emissions that car buyers can expect their new vehicle to produce and forms part of a whole range of designations, listed here in order of decreasing emissions:

TLEV: *Transitional Low Emissions Vehicle*. This is the least stringent emissions standard in California. TLEVs were phased out as of 2004.

LEV: *Low Emissions Vehicle*. All new cars sold in California starting in 2004 have at least a LEV or better emissions rating.

ULEV: *Ultra Low Emissions Vehicle*. ULEVs are 50% cleaner than the average new model year car.

SULEV: *Super Ultra Low Emissions Vehicle*. SULEVs are 90% cleaner than the average new model year car.

PZEV: *Partial Zero Emissions Vehicle*. PZEVs meet SULEV tailpipe emission standards, have zero evaporative emissions and a 15-year per 150,000 mile warranty on its emission control components. No evaporative emissions means that they have fewer emissions while being driven than a typical gasoline car has while idling.

AT PZEV: *Advanced Technology Partial Zero Emissions Vehicle*. AT PZEVs meet the PZEV requirements and have additional "ZEV-like" characteristics. A dedicated compressed natural gas vehicle, or a hybrid vehicle with engine emissions that meet the PZEV standards would be an AT PZEV.

ZEV: *Zero Emissions Vehicle*. ZEVs have zero tailpipe emissions and are 98% cleaner than the average new model year vehicle. These include battery electric vehicles and hydrogen fuel cell vehicles.

Unfortunately, here the term "zero emissions" for electric and hydrogen cars is a fallacy. It ignores the emissions of power plants producing electricity or highly compressed air and CO_2 discharged by hydrogen production out of fuel onboard.

The zero emissions membrane piston engine system (ZEMPES) is an attempt to find a less expensive and more reliable solution to the problem of zero emissions vehicles.

This chapter is aimed at overthrowing the inaction in the ZEMPES demonstration. Converting vehicle fleets in zero emissions operation by ZEMPES does not need the change of fuel supply infrastructure or the radical change of the engines. The gaseous emissions is converted onboard into liquid, which is discharged into fuelling stations and sequestered underground.

The additional mass of equipment onboard ZEMPES is set off by the power increase due to oxygen enrichment, whereas the large mass of a battery, tank with compressed hydrogen, or hydrogen–fuel reformer is the unavoidable penalty for cleanliness.

In view of the large amount of reports describing zero emissions cycles of turbine power, there still exists no description of such cycles for piston engines except as reported by these authors.

7.2 ZEMPES OUTLINE

In this chapter, we continue the line of developing the ZEMPES. It does not contain fuel cell or hydrogen or electrical batteries. It is based on the venerable piston engines which, after more than a hundred years of development, have achieved good properties and reasonable costs. The system does not change the fuel supply infrastructure. The main new element of the system is an ion transport membrane reactor (ITMR), which is close to commercialization due to the achievements of Air Products, Praxair, Norsk-Hydro, and various other companies.

The extended schematic is shown in Figure 7.1(a) with the flows in the ITMR in Figure 7.1(b). Note that in Figure 7.1(a), the reactor is entitled AMR (automotive membrane reactor).

In the scheme presented, there is no supercharger as the power is increased without pressure elevation in the piston engine. The turbocompressor is actually used to feed the ITMR with compressed air, which enhances the oxygen flux. The total system consists of two loops: the main closed loop 1-2-4-5-6-7-1 and an auxiliary loop 18-19-23-24. Fuel enters the mixer at 11, air is taken from the atmosphere at 18, and oxygen is transferred from heated compressed air in the AMR to the mixer at 28 and 29. The combustible mixture at 1 enters a cylinder of VM, is ignited by a spark, and produces mechanical power. The auxiliary turbocompressor supplies compressed air to AMR and gives additional power through clutch KU. The sum is effective power N_e.

Carbon dioxide with dissolved contaminants is deflected from the cycle at 9 to be discharged at a filling station and then sequestered. This part of the system is seen in Figure 2.21, flows 12, 13, and 14.

A 3-D generated design of the ZEMPES for a bus powered by natural gas is shown in Figure 7.2. In this case, the pressure of the natural gas is high enough as is needed for supercritical carbon dioxide. Both gases are separated by a sliding baffle

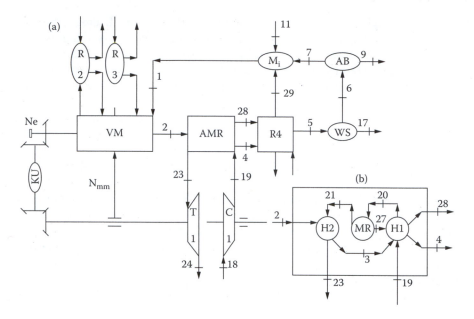

FIGURE 7.1 Schematic of the ZEMPES with oxygen enrichment of an "artificial air" (mixture $O_2 + CO_2$), (Shokotov and Yantovsky, 2006). VM = piston engine, Ne = effective power, R = radiator-cooler, WS = water separator, AB = splitter, Mi = mixer, KU = clutch, N_{mm} = heat flow from mechanical losses of turbocompressor. Numbers reflect the node points.

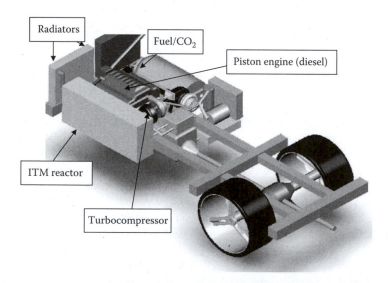

FIGURE 7.2 ZEMPES outlook for a bus on compressed methane (Foy, 2006).

(see Figure 2.21) in one steel cylindrical tank. At the filling station, carbon dioxide is discharged into a central tank, leaving free space for fuel gas.

7.3 HI-OX ZEMPES

In previous papers by McCormick (2003) and Ashley (2005), it was assumed that the fraction of oxygen in the "artificial air" mixture is 0.21, as in air. By increasing this fraction, it is possible to increase the power produced by the combustion. This high-oxygen operation is a new step in the development of the ZEMPES and has been named Hi-Ox ZEMPES. In order to simplify calculations, gas methane is selected as fuel. Calculations for propane and gasoline (petrol) are given later.

Parameters of the selected piston engine are widely used: bore = stroke = 0.092 m, rotational speed 3800 rpm, six cylinders and the compression ratio equal to 16. The volumetric efficiency (filling coefficient) is assumed to be 85%. As the displacement volume is 0.6113 dm^3 and the theoretical amount of reactants in the working volume is 2.5427×10^{-2} mole/cycle, the total flow rates of methane, oxygen and CO_2 are: For complete combustion, $CH_4 + 2O_2 = CO_2 + 2H_2O$, molar masses: $16 + 64 = 44 + 36$. For exact stoichiometry, one mole of methane requires two of oxygen. It is assumed that the total pressure of the mixture is equal at all oxygen fractions. At the points 1, 2, 4, 6, 7, 9, 11, 17, 18, 24, 28 and 29, the pressure is 101.3 kPa, at point 19, it is 506.6 kPa, and at point 23, the pressure is 503.6 kPa. Important assumptions on temperatures are:

$$T_{29} = T_6 = T_7 = T_9 = T_{17} = 313 \text{ K},$$

$$T_{23} = T_2 - 20 \text{ K}, T_4 = T_{28} = T_{19} + 15 \text{ K}.$$

The fraction of oxygen in the oxygen/carbon dioxide mixture varied and the effects on the cycle calculated. Results of these calculations are shown in Table 7.1.

From the table, two important changes are seen when the oxygen fraction is increased:

- The amount of fuel is increased accordingly to maintain the stoichiometry. This leads to an increase in the energy released in each cylinder.
- The amount of recirculated CO_2 is decreased, which leads to a gain in efficiency.

Due to the increase in temperature before the turbine, its power is significantly increased. The increase in power per cylinder allows a reduction in the number of cylinders while maintaining the engine power of 107 kW. At the oxygen fraction of 40%, three cylinders are enough compared to six at the normal fraction, 0.209.

The heat flows in the AMR show a 60-fold increase in heating of the depleted air before the air turbine. In contrast, this heating of incoming air before the membrane reactor is almost equal at all oxygen fractions in order to maintain the air conditions for membrane separation. However, the losses of thermal energy with the discharged depleted air at point 24 have increased up to one third of the fuel thermal power by oxygen fraction 0.5. This heat is lost at a high temperature (1575 K). This is an undesirable situation. Parameters at the node points are shown in Table 7.2. The system is still not optimized.

TABLE 7.1
Results of Calculations for Hi-Ox ZEMPES

	Molar Fraction of Oxygen in Mixture			
Parameter	0.209	0.3	0.4	0.5
Methane flow rate [mole/s]	0.3885	0.5356	0.6844	0.8213
Oxygen flow rate [mole/s]	0.777	1.0712	1.3688	1.6426
Thermal power [kW$_{th}$]	311.71	429.73	549.09	658.91
Indicator power [kW]	112.07	150.84	188.58	222.26
Effective power [kW]	106.57	161.6	215.64	264.12
Turbine power [kW]	68.2	87.07	106	123.2
Compressor power [kW]	41.44	42.94	44.44	45.83
Hot depleted air loss [kW$_{th}$]	99.99	146.84	195.82	242.21
Efficiency	**34.19%**	**37.61%**	**39.27%**	**40.00%**
Temperature T_2 [K]	1,187	1,508	1,831	2,124
Temperature T_3 [K]	1,172	1,297	1,306	1,314
Temperature T_{21} [K]	1,152	1,273	1,273	1,273
Temperature T_{24} [K]	830	1,072	1,319	1,575
Heat Flows in automotive membrane reactor (AMR)				
$h_3 - h_4$ [kW$_{th}$]	130.66	154.78	154.17	153.44
$h_2 - h_3$ [kW$_{th}$]	3.11	45.86	115.58	180.94
$h_2 - h_4$ [kW$_{th}$]	133.77	200.64	269.75	334.38

Source: Shokotov, M. and E. Yantovsky, 2006.

TABLE 7.2
Parameters at Node Points in Hi-Ox ZEMPES

	Molar Fraction of Oxygen in Mixture											
	0.209			0.30			0.40			0.50		
Point	T K	Flow Mole/s	H kW	T K	Flow Mole/s	H kW	T K	Flow Mole/s	H kW	T K	Flow Mole/s	H kW
1	311	4.1	49.4	310	4.1	47.3	310	4.1	45.2	309	4.1	43.3
2	1,187	4.1	220.8	1,508	4.1	286.2	1,831	4.1	353.8	2,124	4.1	417
4	507	4.1	87	507	4.1	85.5	507	4.1	84	507	4.1	82.6
6	313	3.29	44.5	313	3.03	40.5	313	2.73	36.6	313	2.46	32.9
7	313	2.94	39.3	313	2.25	33.4	313	2.05	27.4	313	1.64	21.9
9	313	0.39	5.19	313	0.53	7.15	313	0.68	9.15	313	0.82	11
11	293	0.39	3.22	293	0.53	4.43	293	0.68	5.67	293	0.82	6.8
17	313	0.77	2.34	313	1.07	3.23	313	1.37	4.12	313	1.64	4.9
18	293	7.08	5.87	293	7.33	60.85	293	7.6	63	293	7.83	65
19	492	7.08	100.2	492	7.33	103.8	492	7.6	107.4	492	7.83	110.8
23	1,167	6.3	222.6	1,488	6.26	288.8	1,811	6.22	357.2	2,104	7.83	421.2
24	830	6.3	154.4	1,072	6.26	201.7	1,319	6.22	251.2	1,575	6.19	298
28	507	0.77	11.3	507	1.07	15.63	507	1.37	19.97	507	1.64	23.9
29	313	0.77	6.9	313	1.07	9.5	313	1.37	12.15	313	1.64	14.5

Source: Shokotov, M. and E. Yantovsky, 2006.

7.4 ADDITION OF THERMOCHEMICAL RECUPERATION (TCR)

Bottoming cycles to recover the thermal losses from exhaust gases of gas turbines and piston engines are well known. Some such additions to the ZEMPES were examined (Shokotov et al., 2005). However, addition of bottoming cycles with rotating parts makes the total system very complicated and impractical. A static system, similar to an additional heat exchanger, would be better. Such a device is the thermochemical recuperator, or TCR, see also Section 9.3. The TCR converts methane and water to syngas, using the endothermic reaction:

$$CH_4 + H_2O \Rightarrow CO + 3H_2 \tag{7.1}$$

We have calculated the change in engine parameters if a TCR is installed after the air turbine. The new fuel, a combustible gas consisting of $CH_4 + H_2O + CO + H_2$, is cooled in a radiator-cooler before entering the cylinder. The molar fractions are: 0.05556 CH_4, 0.05556 H_2O, 0.2222 CO, and 0.6666 H_2 when the reaction begins at 524.7 K and ends at 588.4 K.

These temperatures look rather low with respect to common practice of methane conversion with steam (about 1000 K). They have been calculated exactly using the Gibbs energy change equation (Hoverton, 1964) for 80% conversion:

$$G_0 = 232,977 + 109.28 \cdot T - 68.18T \cdot \ln T + 40.35 \cdot 10^{-3} \cdot T^2 - 3.67 \cdot 10^{-6} \cdot T^3 \tag{7.2}$$

These temperatures, if necessary, could be increased up to that of the exhaust gases less a heat transfer drop. Heating of incoming reactants is the main process of energy conservation in the TCR. The air after flowing through the turbine is cooled before being discharged into the atmosphere. The power generated and temperatures at some points when a TCR is incorporated into the system are presented in Table 7.3

TABLE 7.3
Results of Calculations for Hi-Ox ZEMPES with TCR

Parameter	Molar Fraction of Oxygen in Mixture			
	0.209	0.3	0.4	0.5
T_2 (after piston engine), [K]	1,322	1,688	2,063	2,375
T_{24} (before TCR), [K]	1,108	1,359	1,622	1,840
T_{25} (after TCR), [K]	816	970	1,149	1,295
Pressure Ratio	**2.18**	**2.796**	**3,338**	**3.778**
Effective power, [kW]	—	133.54	161.0	185.1
Efficiency	**—**	**41.62%**	**42.03%**	**42.70%**
Increased Pressure Ratio	**—**	**3**	**5**	**9**
Oxygen partial pressure ratio	—	3.219	4.49	7.15
Effective power, [kW]	—	135.06	170.33	204.6
Efficiency	**—**	**42.09%**	**44.46%**	**47.19%**

Source: Shokotov, M. and E. Yantovsky, 2006.

TABLE 7.4
Possible Coatings for Adiabatic Engines

Material	Flexure Strength [MPa]	Density [g/cm³]	Young's Modulus[a] [GPa]	Thermal Expansion[b] $10^6 \times$ [1/K]	Thermal Conductivity [W/mK]
Si_3N_4	300	3.1	300	3.2	12
SiC	450	3.15	400	4.5	40
Al. Magn. Silicate	20	2.2	12	0.6	1
ZrO_2	300	5.7	200	9.8	2.5
Al_2O_3 /TiO_2	20	3.2	23	3	2

[a] At the temperature 1260 K.
[b] In temperature range 300 to 1260 K.

at pressure ratios 2.18 to 9. The highest efficiency is 47.19% at an oxygen fraction of 0.5 and a pressure ratio equal to 9.

Calculations show the feasibility of the zero emissions piston engine system (called HiOx-ZEMPES), using an ordinary fuel (methane), with increased amounts of fuel and oxygen in the cycle by combustion in an *artificial air* which is a mixture of oxygen from the membrane reactor and recirculated carbon dioxide. At oxygen molar fraction 0.5 and pressure ratio 9 in the air turbocompressor, the system efficiency with thermochemical recuperation is calculated as 47%.

Here, the efficiency reached 47%, which might be encouraging for fuel-cell engines. Such a high efficiency gain is possible because the temperature of the exhausted depleted air after the TCR is much less (1295 K) than after the turbine in Table 7.1 (1575 K). By further exploration of this system, it might be possible to increase efficiency further by cooling exhausted air in TCR.

To withstand the rather high temperatures in cylinders and valves, the great progress in *adiabatic* engines might be used. They are also called "low heat rejection engines (LHRE)" (Leidel, 1997, and Jaichandar, 2003). Layers of refractory ceramics, e.g., stabilized zirconia or sintered silicon nitride have been used successfully. Table 7.4 lists possible coatings.

Both of the reactors, ITMR and TCR, operate at significant temperatures. They require heating at start-up, which takes some time. Both reactors should be switched off during start-up, admitting starting emissions. However, these emissions are very small with respect to the emissions avoided during continuous operation, when the system is switched on and working.

7.5 MEMBRANE REACTOR FOR PISTON ENGINES

In some versions of the ZEMPES, the onboard membrane reactor requires sweeping of oxygen from the membrane, i.e., air is on one side of the membrane and carbon dioxide on the other. Oxygen that passes through the membranes is swept away by the carbon dioxide. Such a reactor, shown in Figure 7.3, has been developed by Norsk Hydro for the AZEP project (Sundkvist et al., 2007). Due to the crucial importance of the oxygen supply by a monolith reactor for automotive use, we reproduce the data in some detail.

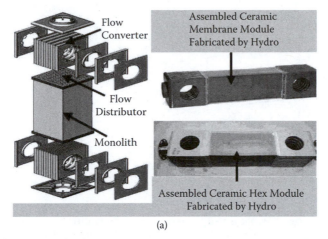

(a)

FIGURE 7.3a ITMR for producing mixture of oxygen and carbon dioxide, developed by Norsk Hydro for the AZEP project (Sundkvist et al., 2007).

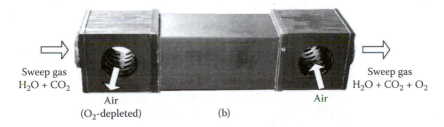

(b)

FIGURE 7.3b The module of Monolith reactor, carefully tested (Julsrud et al., 2006).

Figure 7.3 to Figure 7.7 are taken from a complex investigation (Sundkvist et al., 2007) of a membrane reactor by Siemens (Sweden), Norsk Hydro (Norway) and ALSTOM (Switzerland). In Figure 7.3, we see the structure of the reactor and heat exchangers. They are rather robust and strong enough to be used onboard a vehicle.

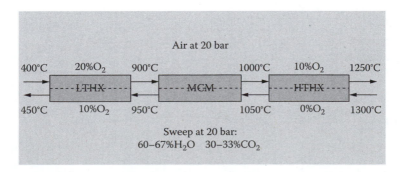

FIGURE 7.4 Schematics of flows in reactor MCM and neighboring heat exchangers for the AZEP project (Sundkvist et al., 2007), similar to ZEMPES.

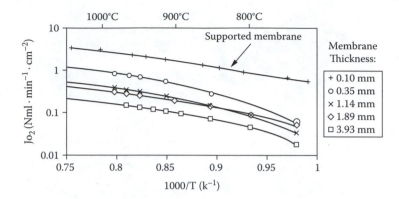

FIGURE 7.5 Oxygen flux, measured by pressure 1 bar and different membrane thickness. The unit of the oxygen flux on Y axis is equivalent to 0.238 g/m² × s. The upper curve reflects the reactor case (Julsrud et al., 2006).

The recommended temperatures in the system reactor-heat exchangers for a nominal condition are seen in Figure 7.4. In Figure 7.5, we see the unique experimental data on oxygen flux, never before published. There exist a lot of published measurements of oxygen flux through a sample similar to a coin. It was a concern of deviation in case of the test of the whole module. Actually, the difference is negligible. Figure 7.6 vividly shows the influence of the temperature on the reactor work. Finally, Figure 7.7 illustrates the work of the heat exchanger by increase and decrease of specific heat flow. The sweep gas coming into the reactor is rather hot,

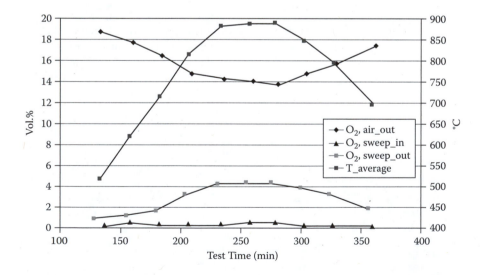

FIGURE 7.6 Oxygen concentration and temperature during a test of the module. Temperature increase is accompanied by oxygen in sweep gas increase.

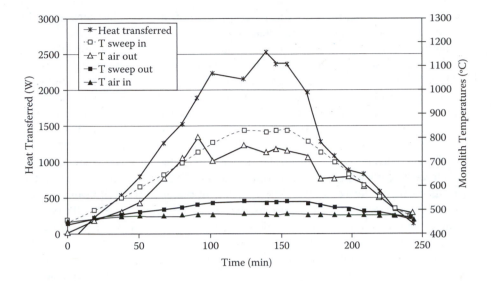

FIGURE 7.7 Increase and decrease of temperatures in the heat exchanger versus time. One of the obstacles of the use in transient regimes: thermal inertia.

along with outgoing air, an oxygen donor. The incoming air after compressor and outgoing sweep gas are relatively cold.

7.6 ZERO EMISSIONS TURBODIESEL

In order of the soonest demonstration, the diesel engine is selected because it is the work horse of American economics with very dirty emissions. In contrast to a popular slogan of proponents of fuel cells, "New diesels will help bridge the gap to future fuel-cell vehicles" (Ashley, 2007), which need costly sulphur-free fuel, in ZEMPES the sulphur oxides would be dissolved in liquid carbon dioxide and stored together underground. That is why the large sulphur removal units and exhaust treatment systems with urea tanks and ammonia absorbers are not needed. The crucial thing is that those cumbersome systems admit the carbon dioxide emissions. They can prevent the poisonous gases and particulate matter emission only, whereas ZEMPES prevents all the combustion-born emissions.

The schematics are similar to previous ZEMPES versions (Yantovsky, 2003, and Shokotov et al., 2006). The new element is the afterburner.

In turbodiesel, there are three sources of thermal energy which might be converted into power:

1. Fuel combustion in cylinder
2. Fuel combustion in the afterburner
3. Hot flue gases

The crucial feature is the efficiency of the turbodiesel. It is defined as the ratio of effective mechanical power (of piston engine + turbine) to the thermal energy of both fuel flows. The thermal energy of flue gases originates from the same fuels. It should be used in the turbine to maximize efficiency.

In this section, only steady-state operation is considered. Any transient regimes are still under discussion. For that matter, what should be known is the thermal inertia of ITM reactor. In starting regimes, it probably would first be a switched-on diesel with emissions to atmosphere (admitted for the short time), and then attached to the turbo-compressor and membrane reactor, which should be heated up to 900°C (Figure 7.8).

The working process consists of the two well-known cycles: closed diesel cycle 1-2-4-5-6-7-1 and open Brayton cycle 18-19-20-21-22-23-24. Let us start from the diesel cycle. The mixture of oxygen and carbon dioxide (artificial air, where nitrogen is replaced by carbon dioxide) is exposed to suction. After compression, it is injected with diesel fuel. Expansion gives power, then displacement of much of the oxygen contained gases, driving them into the afterburner VK. Here, almost all of the oxygen in the combustor is used to burn the additional fuel (gasoline) and to heat air before entering the turbine.

After AK, the amount of oxygen is nil and flows in node points 3, 4, 5 are identical. After cooling, the water is liquid but CO_2 is gaseous. They are separated in WS, water is deflected and it might be used elsewhere or to increase turbine power. Almost dry, CO_2 is split into AB. A minor part, exactly equal to combustion born CO_2, is liquified onboard by cooling, compression, and more cooling, and remains onboard under pressure of about 100 bars to be discharged into a central tank at a fuel

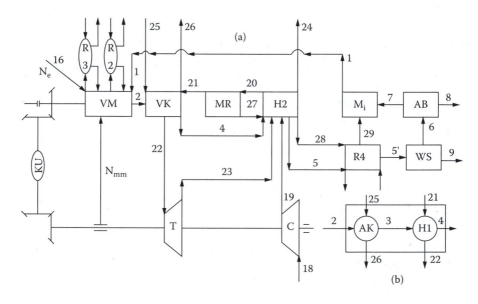

FIGURE 7.8 Schematics of turbodiesel (Shokotov and Yantovsky, 2007). (a) R2 and R3 = oil and water radiators, VM = diesel piston engine, VK = afterburner, which consists of a combustor AK and heat exchanger H1, (b) MR = membrane reactor for oxygen separation, H2 = multipurpose heat exchanger, Mi = mixer, AB = gas splitter, WS = water separator, R4 = cooler, N_{mm} = heat flow of mechanical losses in turbocompressor, KU = clutch, T = turbine, C = compressor.

filling station. The major part of CO_2 goes back to be mixed with oxygen and to form artificial air as an oxydizer in the diesel cycle.

The open (Brayton) cycle starts from ambient air in 18. After filter and compressor, it is heated in H2 and enters the membrane reactor MR, where approximately 60% of oxygen is penetrated through the membrane wall into the CO_2 flow. The rest of the air is heated in the afterburner and expands in turbine T, giving additional power through the reductor before KU. The reductor is essential as the piston engine shaft rotates by 4,000 rpm, whereas the turbocompressor shaft frequency is approximately 60,000 rpm. The depleted air after turbine is absolutely harmless, and contains no combustion products.

7.7 MEMBRANE REACTOR FOR TURBODIESEL

Reactors are on the verge of practical use due to the achievements of Air Products and Norsk Hydro (Julsrud, 2006); see Figure 7.2 in Section 7.5. The recent available figures of measured oxygen flux (Sundkvist, 2007) through membranes of 1.4 mm thickness, made of perovskite $Ba_x Sr_{1-x} Co_y Fe_{1-y} O_{3-\delta}$ by $T = 850°C$ are definitely about oxygen flux of 8 ml/cm^2 × min = 1.9 g/m^2 × s. Important work for reactor design has been done in ECN, the Netherlands by Vente et al. (2005).

Discussions of three different configurations of membranes in a reactor of a rather big size for practical applications have been underway. They admit the oxygen flux as 10 ml/cm^2 × min, which is near to measured data. In our calculations, the modest figure of $j_{O2} = 1.9$ g/m^2 × s is used.

For the monolithic type of reactor assembled of tubes of an outer diameter of 200 mm with channels of 1.5 mm and wall thickness of 0.5 mm (Figure 7.10), the important parameter was calculated as the ratio of active surface to volume as 400 m^2/m^3. It is indicated by an arrow in Figure 7.9.

For power 217 kW turbodiesel needs 1.102 mole/s of O_2 or 35.26 g/s (Table 7.5, point 29). Assuming the surface/volume ratio as 400 m^2/m^3 and $j_{O2} = 1.9$ g/m^2 × s, we need a surface of 35.264/1.9 = 18.56 m^2 or the volume of active membranes as 0.0464 m^3. By 0.2 m diameter, the length of tubular reactor is 0.0464/0.0314 = 1.477 m. It might be in two modules of 0.75 m each. By specific weight for heaviest ceramics (zirconia) of 6000 kg/m^3 and ceramic volume 0.64 of active volume (due to channels), its mass is 0.0464 × 0.64 × 6,000 =178.17 kg. Steel shell mass is: 1.5 × 3.14 × 0.2 × 0.002 × 8000 = 15.06 kg. The total mass of ITMR is 178.75 + 15.06 = 193.8 kg. As the mass of prototype diesel engine is 219 kg, the mass of ITMR is near 88% of its value.

Here exists the principal problem of ZEMPES for vehicles—extra weight. There are three causes of additional weight: membrane reactor, high temperature heat exchanger H1, and liquid carbon dioxide onboard.

The mass of the prototype engine is 219 kg. Imagine the fuel mass as 51 kg, so total mass is 270 kg, by nominal power 52 kW, hence, specific weight of a prototype is 270/52 = 5.19 kg/kW.

The question is zero emissions operation needs some additional equipment. Is it possible to increase power of turbodiesel to such an extent as to conserve the specific weight?

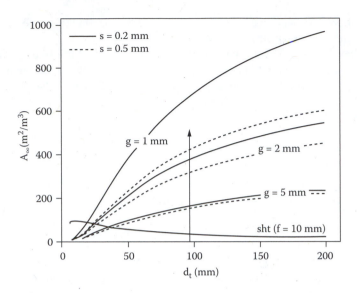

FIGURE 7.9 Geometrical properties of monolith ITM reactors of cylinder shape, calculated in (Sundkvist et al., 2007). Arrow shows the design point, selected for turbodiesel. Selected parameters; see also in Figure 7.10.

The next tables give the positive answer to this question due to increase of the oxygen fraction in the cylinder charge. If the mass of produced and captured CO_2 is three times that of the fuel mass, it means 153 kg, ITMR mass of 193 kg, and heat exchanger H1 of 50 kg. The specific weight of turbodiesel is:

$$(219 + 153 + 193.8 + 50)/217 = 2.837 \text{ kg/kW}$$

which is about two times less than that of the prototype.

7.8   NUMERIC EXAMPLE

The well-known diesel engine Ford-425 has been selected for calculations as a prototype. Manufacturer data are as follows: bore = 93.67 mm, stroke 90.54 mm, 4 cylinders, rotational speed 4000 rev/min, volume 2.5 dm³, compression ratio 19.02, effective power 52 kW; see Table 7.6.

We have calculated the engine efficiencies: indicated 0.478, mechanical 0.737, and effective 0.352.

In Table 7.6, the first column relates to the prototype (diesel Ford 425) and the other 4 columns describe the turbodiesel. The table reveals the major role of the turbine, due to 4 times the power increase and significant efficiency gain with respect to the prototype.

Efficiency gain from 35.22% up to 48.2% compensates for the need to carry rather heavy equipment onboard.

TABLE 7.5
Parameters of Turbodiesel in Node Points

		Case 1			Case 2			Case 3			Case 4		
N	P kPa	T	M	H	T	M	H	T	M	H	T	M	H
1	2	3	4	5	6	7	8	9	10	11	12	13	14
1	101.3	308.0	2.755	32.553	308.0	2.755	31.339	308.0	2.7547	32.553	308.0	2.7547	31.339
2	101.3	1,249	2.941	165.02	1551	3.003	212.14	980.8	2.8852	121.47	1205	2.9288	151.51
3	101.3	1,780	3.023	260.91	2122	3.112	339.99	1939	3.0283	289.27	2365	3.1194	375.25
4	101.3	1,283	3.023	178.12	1283	3.112	181.25	1283	3.0283	178.29	1283	3.1194	181.44
5	101.3	527.0	3.023	66.98	527.0	3.112	68.11	527.0	3.0283	67.04	527.0	3.1194	68.18
6	101.3	308.0	2.49	32.722	308.0	2.402	31.563	308.0	2.4836	32.638	308.0	2.3932	31.451
7	101.3	308.0	1.928	25.34	308.0	1.653	21.72	308.0	1.9283	25.34	308.0	1.6528	21.72
8	101.3	308.0	0.5617	7.382	308.0	0.7490	9.8426	308.0	0.5553	7.298	308.0	0.7404	9.730
9	101.3	308.0	0.5329	1.4059	308.0	0.7106	1.8745	308.0	0.5447	1.437	308.0	0.7262	1.916
18	101.3	288.0	13.377	109.01	288.0	17.718	144.39	288.0	14.841	120.94	288.0	19.542	159.25
19	608.0	511.8	13.377	197.1	511.8	17.718	261.06	511.8	14.841	218.67	511.8	19.542	287.93
20	608.0	1,273	1.3377	519.06	1,273	17.718	687.5	1,273	14.841	575.87	1,273	19.542	758.28
21	608.0	1,273	12.550	486.99	1,273	16.616	644.75	1,273	14.014	543.8	1,273	18.44	715.51
22	608.0	1,469	12.550	569.79	1,555	16.616	803.49	1,507	14.014	654.78	1,507	18.44	909.32
23	101.3	1,017	12.550	381.96	1,081	16.616	539.71	1,046	14.014	439.32	1,101	18.44	611.2
24	101.3	527.0	12.550	190.66	527.0	16.616	252.42	527.0	14.014	212.89	527.0	18.44	280.13
27	101.3	1273	0.8264	32.067	1,273	1.102	42.761	1,273	0.8264	32.067	1,273	1.102	42.76
28	101.3	527.0	0.8264	12.554	527.0	1.102	16.741	527.0	0.8264	12.554	527.0	1.102	16.74
29	101.3	308.0	0.8264	7.214	308.0	1.102	9.619	308.0	0.8264	7.214	308.0	1.102	9.619

Note: P = Pressure, kPa; T = Temperature, K; M = Mass Flow, Mole/s; H = Enthalpy. Flow, kJ/s.

TABLE 7.6
Calculated Power and Efficiency of Turbodiesel

Oxygen Fraction Q	0.209 (Air)	0.3	0.4	0.3	0.4
Excess coefficient, α	1.7	1.4	1.4	2.0	2.0
Diesel fuel flow, g/s	3.473	5.684	7.579	3.979	5.305
Thermal power, kW	147.4	241.3	321.7	168.9	225.2
Added fuel, g/s	0	2.180	2.906	3.814	5.086
Added thermal power, kW	0	96.50	128.7	168.9	225.2
Total thermal power, kW	147.4	337.8	450.3	337.8	450.3
Diesel power, kW	52	66.68	90.65	45.14	64.50
Turbine power, kW	0	85.01	126.4	100.8	146.1
Total power, kW	**52**	**151.69**	**217.08**	**145.98**	**210.57**
Efficiency	**0.3522**	**0.4491**	**0.4820**	**0.4322**	**0.4676**

Source: Shokotov, M. et al., 2005.

In ZEMPES, the use of a membrane reactor (see Figure 7.10) gives an ability to govern the oxygen flow rate and quite independently, change both of the governing quantities: oxygen excess coefficient α (ratio of actual flow rate to the stoichiometric one) and oxygen fraction in a mixture with CO_2 (different amount of O_2 in artificial air). Fixed values only are used in the tables (1.4 or 2.0 and 0.3 or 0.4).

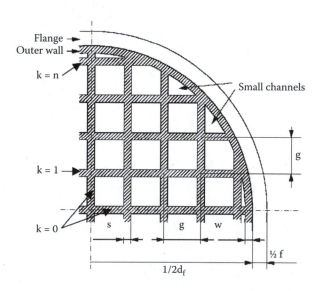

FIGURE 7.10 Cross-section of one quarter of a multichannel monolith ITM reactor selected for the turbodiesel: $d_f = 200$ mm, $s = 0.5$ mm, $g = 2$ mm, $f = 4$ mm. (In Figure 7.8 for this case, the surface/volume ratio is 400).

Now, we may compare the three main characteristics of the membrane reactor, tested in Section 7.7, and calculated here.

The test calculation results referred to $1 m^3$ of the reactor volume are:

Surface/volume ratio of active part	500 m^2/m^3	400 m^2/m^3
Oxygen flow	37 mole/ $m^3 \times s$ = 1,184 kg/$m^3 \times s$	760 g/$m^3 \times s$
Thermal power	15 MW/m^3	217 kW/0.0464 m^3 = 4.67 MW/m^3

In the last column, the reflected calculation is with rather modest figures with respect to results achieved in experiments. It indicates a good perspective for further improvement of turbodiesel.

7.9 HIGH-TEMPERATURE HEAT EXCHANGER FOR TURBODIESEL

Afterburner AK (see Figure 7.8) emits very hot gas to the heat exchanger H1. Its energy is used to heat the compressed depleted air. In Table 7.2, the entrance temperature is observed to be 2122 K (1849°C), which is too high to use in any heat exchanger made even of the best known steel.

The only solution we see is to use the well-known material SiC, silicon carbide. It is quite stable even in oxidizing flow by such temperatures and is widely used for manufacturing by extrusion. SiC is stable against thermal and mechanical shocks and has high thermal conductivity.

Good perspectives of ceramic heat exchangers for microturbines, as in our case, are forecasted by McDonald (2003) in a comprehensive review.

TABLE 7.7
Parameters of the Heat Exchanger H1

Quantity	Unit	Hot Line (CO_2)	Cold Line (Air)
Mass flow	mole/s, (g/s)	3.112 (130)	16.6 (481)
Temperature in	K	2,122	1,273
Temperature out	K	283	1 554
Enthalpy in	kW	340.0	644.7
Enthalpy out	kW	181.3	803.5
Viscosity	$10^7 \times Pa \times s$	655	674.4
Density	kg/m^3	1.56	1.0
Thermal conductivity	W/m $\times$ K	0.121	0.090
Mean velocity	m/s	10	25
Reynolds Number	—	476	740
Nusselt Number	—	1.8	2.0
Heat transfer coefficient	W/$m^2 \times$ K	109	128

Our heat exchanger is similar to the ones of monolith type HEX, made by Norsk Hydro; see Figure 7.3. Because it is an important part of the turbodiesel, the detailed data from a preliminary calculation to guess its mass and dimensions are given in Table 7.5.

An ideal thermal insulation without any loss of energy from H1 was assumed. The radiation heat exchange has been neglected as well as conductive thermal resistance in SiC due to high thermal conductivity (490 W/m K).

The square channel of 2 mm side is assumed in both lines. Overall heat transfer coefficient is 58.8 W/m^{2}K. Mean logarithmic temperature drop is 139 K. Mean specific heat current is $139 \times 58.8 = 8{,}173$ W/m^2. The heat transfer surface is $158.74/8.173 = 19.4$ m^2.

As a guarantee that such a device might be manufactured of SiC by extrusion, one may have a look at the CeraMem Technology brief (2006). A cylinder device with passageway of 2 mm, with the active surface 10.7 m^2, diameter 142 mm and length 864 mm has the mass of about 25 kg. For our example of a turbodiesel, two such cylinders are enough, with a total mass of 50 kg. It is not prohibitive even for a car.

Calculations show the possible increase of power from 52 to 217 kW and efficiency from 35 up to 48% by conversion of a car diesel to zero emissions turbodiesel with a membrane oxygen reactor. The additional mass of a membrane reactor is about the mass of a prototype engine. The total mass of the system in a turbodiesel increased to two times less than power. It justifies further research on a turbodiesel project for consideration in the near future.

7.10 ECONOMICS OF ZEMPES ON DIFFERENT FUELS

In spite of the lack of some important figures, we try here to evaluate some possible economic benefits of the described system. The least known at the moment is the price of the ITM reactor; it was assumed preliminary.

The system consists of well-developed devices such as the ordinary piston engine, the turbocompressor, and heat exchangers. The only new element is the ion transport membrane reactor, which, after much work by industrial companies, is very near to commercialization (Armstrong et al., 2003; Sundkvist et al., 2004; Selimovich, 2005). In the economic evaluations presented, the current costs are known with available accuracy of fuel prices.

The schematics are in Figure 7.1a with the flowsheet of a membrane reactor (MR) in Figure 7.1b. Note, at Figure 1a, the reactor for oxygen is entitled AMR (automotive membrane reactor).

As in previous chapters, the components are listed as VM = piston engine, N$_e$ = effective power, R = radiator-cooler, WS = water separator, AB = splitter, Mi = mixer, KU = clutch, N$_{mm}$ = heat flow from mechanical losses of turbocompressor. Numbers reflect the node points.

In the discussed scheme, there is no supercharger as the charge is increased without pressure elevation. Turbocompressor is actually used to feed the ITMR by compressed air which enhances the oxygen flux only.

The total system consists of the two loops: the main closed loop 1-2-4-5-6-7-1 and an auxiliary one 18-19-23-24. Fuel comes to the mixer in 11, air is taken from the atmosphere in 18, oxygen goes from the heated compressed air in AMR to the mixer in 28, 29. The combustible mixture in 1 enters a cylinder of VM, is ignited by spark, and produces mechanical power. The auxiliary turbocompressor supplies compressed air to the AMR and gives additional power through clutch KU. The sum is effective power N_e.

Carbon dioxide with dissolved contaminants is deflected from the cycle in 9 to be discharged at the filling station and then sequestered.

The membrane reactor AMR is presented in Figure 7.10 in some detail, including monolith type reactor MR itself and similar construction heat exchangers (Sundkvist et al., 2004; Selimovich, 2005).

After compression in C1, air is heated by flue gases 3 in H1 up to 800–1000°C and goes to the reactor MR. The oxygen partial pressure ratio across the membrane is maintained over 3. Pure oxygen 28 to 29 travels via R4 to the mixer Mi, then the combustible mixture enters the piston engine VM.

Depleted air 21 is heated in H2 by hot flue gases and expanded in the air turbine T1 to ambient pressure.

The pressure ratio in the cylinder is restricted by temperature after compression. In ZEMPES, where the inert gas is CO_2 instead of N_2, it is possible to increase the pressure ratio from about 8 to about 16 to 17 by means of a minor redesign of the cylinder. Calculations show the ability to restrict temperature by the admissible 600 to 700 K.

As a piston engine, the selection was the Volkswagen engine with 4 cylinders in a row, stroke 86.4 mm, bore 79.5 mm, volume ratio 8.3, frequency 5000 rpm, and effective power 48 kW. Results are seen in Table 7.9.

Table 7.8 gives the total picture of the thermodynamic parameters of one regime (not optimized).

Rather high temperatures of more than 1000 K of air after turbine even by modest oxygen enrichment of 0.3 suggest preventing such exergy losses by further system elaboration. By an oxygen fraction of 0.5, this temperature exceeds even 1500 K. But such enrichment might not be recommended. There appear to be many engineering problems, because after expansion in the cylinder, the flue gases are at more than 2100 K and after combustion, more than 3500 K. Here, some warnings are needed. Calculations did not take into account the dissociation of gases by high temperatures. Actually, such temperatures will not be achieved. But such high-temperature region is out of engine practice, the combustion process is to be quick, perhaps the assumed equilibrium is not achieved, and real composition of combustion gases is variable.

Here, the cost calculations were the price of gasoline at 1.24 €/liter, propane at 0.639 €/liter and natural gas (methane) at 0.829 €/kg, as it was in Germany in January 2006.

It is apparent the system with methane fuel gives maximum saving, as it is well known from the current practice of using natural gas for vehicles. It is attractive for a decentralized power production by small units, which are zero emissions. Therefore, it might be used in densely populated areas. There is little doubt there might be a co-generating of power and heat.

Even using gasoline, the fuel economy is very significant, which contradicts a common view that a zero emissions power unit should be less efficient than conventional ones due to the need to produce oxygen and liquify carbon dioxide. In many

TABLE 7.8
Parameters in Node Points for HiOx-ZEMPES for Gasoline (B), Propane and Methane as Fuels

Fuel	O_f	Par.	Unit	1	2	4	9	11	17	19	23	24	28
B	0.3	T	K	309	1,540	535	313	293	313	520	1,520	1,055	535
		M	mole/s	2.20	2.33	2.33	0.424	0.0525	0.449	4.34	3.69	3.69	0.646
		H	kW	29.9	172.6	53.4	5.66	3.98	1.35	65.1	174.3	117.0	9.98
	0.5	T	K	306	2,165	535	313	293	313	520	2,145	1,525	535
		M	mole/s	2.217	2.25	2.25	0.696	0.086	0.737	4.93	3.86	3.86	1.06
		H	kW	30.24	266	54.7	9.30	6.54	2.22	73.9	268.8	183.0	16.45
C_3H_8	0.3	T	K	310	1,555	535	313	293	313	520	1,535	1,066	535
		M	mole/s	2.26	2.38	2.38	0.383	0.126	0.511	4.33	3.69	3.69	0.639
		H	kW	27.58	174.1	53.3	5.12	1.97	1.53	64.9	175.8	118.0	9.97
	0.5	T	K	309	2,179	535	313	293	313	520	2,159	1,536	535
		M	mole/s	2.26	2.47	2.47	0.618	0.206	0.824	4.82	3.79	3.79	1.03
		H	kW	26.1	263	53.6	8.26	3.23	2.48	72.3	265.8	181.0	15.9
CH_4	0.3	T	K	310	1,524	535	313	293	313	520	1,504	1,044	535
		M	mole/s	2.26	2.26	2.26	0.294	0.294	0.589	4.04	3.45	3.45	0.53
		H	kW	26.0	159.3	49.86	3.93	2.44	1.77	60.6	150.5	108.0	9.09
	0.5	T	K	309	2,117	535	313	293	313	520	2,097	1,488	535
		M	mole/s	2.26	2.26	2.26	0.453	0.453	0.906	4.33	3.42	3.42	0.906
		H	kW	23.9	229.2	48.3	6.06	3.75	2.73	64.9	231.7	157.6	14.0

Note: O_f = Oxygen fraction; Par. = parameter.

papers, it is indicated that the energy penalty is 8 to 10 percentage points for cryogenic oxygen and 2 to 4 percentage points for membrane oxygen. In our calculations, it is evident that there is a rise of efficiency from 27% in prototype to up to 40% in ZEMPES due to reasonable forcing by oxygen enrichment and additional power from depleted air turbine.

We should stress that the figures of very high oxygen enrichment 0.5 are included in order to give a perspective. The maximum pressure of 178 bars along with temperature of 3500 K is not achieved in current practice. For realistic economic evaluations, we use modest oxygen enrichment 0.3 (it is 1.5 times more than in air): fuel gasoline, its specific consumption in prototype 300 g/kWh, and in ZEMPES 220 g/kWh by efficiency 0.368, fuel price 1.77 €/kg.

For vehicle applications, the size of the membrane reactor is of crucial importance. Let us evaluate it for ordinary fuel: gasoline, which is modeled as 1-hexene C_6H_{12} or for stoichiometry calculations CH_2. The combustion equation is:

$$CH_2 + 1.5\,O_2 \;\Rightarrow\; CO_2 + H_2O + 45 \text{ kJ/g } CH_2 \text{ or } 13 \text{ kJ/g } O_2$$

Assuming the oxygen flux of 1 g/m² × s, we get the energy current through the membrane equal to 13 kW/m².

In membrane reactors of monolith or tubular type, the modest quantity of specific surface of 100 m²/m³ is easily achievable (Sundkvist, et al., 2004; Selimovich, 2005). The needed thermal power for 100 kW ZEMPES engine by efficiency 0.368 is 272 kW_{th}. Such thermal power is given by the membrane surface of 272/13 = 21 m².

The active volume for such surface is 21/100 = 0.21 m³, or 0.6 × 0.6 × 0.6 m. The volume of additional heat exchangers depends on the construction but is almost equal to that of a reactor. It means the total needed volume is about 0.6 m³, which seems to be reasonable for such a vehicle. For steady-state power production or a bus, such size is evidently admissible.

The first cost (investment) of ZEMPES is evidently higher than that of a prototype, a conventional piston engine, due to adjustment to increased temperature and pressure along with new elements such as a membrane reactor, heat exchangers, and turbocompressor. However, the current cost may be less due to fuel saving. The main question of economics is the payback time: how long should we wait for the return of extra investment?

Imagine the extra cost of ZEMPES with respect to a prototype of the same 100 kW, having emissions, is 10,000 € (say, 30,000 for ZEMPES with respect to 20,000 of prototype), the time of use by full capacity is 1,000 hours annually, hence, 100,000 kWh per year. Annual gasoline consumption is 30 t for a prototype and 22 t for ZEMPES, fuel economy is 8 t/year. If the fuel price is 1.77 €/kg, the annual saving is 1.77 × 8,000 = 14,160 €. This means the payback time is only 8.5 months, less than a year. The example is illustrative only because the real cost of the membrane reactor is still unknown.

Calculations show the evident possibility to create a zero emissions piston engine unit for a vehicle or decentralized power production, which might use gasoline, propane, or natural gas with significant reduction of fuel consumption and current costs. For a modest oxygen enrichment 0.3 (instead of 0.209 in air), by attainable maximum

TABLE 7.9

Power, Efficiency, Fuel Consumption and Cost of Energy (COE)

Quantity	Unit	Prototype B	Prototype C_3H_8	Prototype CH_4	ZEMPES, 0.3 B	ZEMPES, 0.3 C_3H_8	ZEMPES, 0.3 CH_4	ZEMPES, 0.5 B	ZEMPES, 0.5 C_3H_8	ZEMPES, 0.5 CH_4
Compress. ratio	—	8.3	8.5	8.5	16.0	17.0	17.0	18.0	19.0	19.0
Compress. temp.	K	683	701	755	590	655	709	592	663	756
Thermal power	kW	177	180	167	263	261	236	431	421	363
Indicated power	kW	69.0	70.0	64.0	92.9	87.7	78.7	150	139	120
Indicated efficiency	%	38.9	38.8	38.3	35.3	33.6	33.3	34.8	33.1	33.0
Mean pressure	kPa	965.0	981.0	899.0	1,300	1,288	1,102	2,099	1,953	1,680
Effective power	kW	48	49	43	96	92	82	176	166	140
Effective efficiency	%	27.1	27.2	25.9	36.8	35.3	34.7	40.9	39.4	38.5
Max. temperature	K	2,800	2,780	2,730	2,640	2,650	2,610	3,580	3,570	3,500
Max. pressure	bar	55.9	58.6	55.5	110.0	113.0	105.9	178.7	177.8	159.7
Fuel consumption	kg/h	14.5	14.0	12.0	21.5	20.2	16.9	35.4	32.6	26.1
Spec. fuel cons.	g/kWh	303	285	277	223	220	207	200	197	187
CO_2 emission	kg/h	45.3	42.0	33.0	0	0	0	0	0	0
COE	€c/kWh	53.6	35.7	23.0	39.4	27.6	17.1	35.5	24.7	15.5
Saving	%	—	33.3	57.2	26.4	48.5	68.0	33.8	54.0	71.1

temperature and pressure, for 100 kW capacity, the efficiency reaches 0.368 and fuel cost economy 26% on gasoline and 68% if going from gasoline to natural gas. The described HiOx-ZEMPES system deserves immediate consideration.

7.11   PISTON ENGINE WITH PRESSURE SWING ADSORPTION OXYGEN REACTOR

The change in this section is the use of a PSA (pressure swing adsorption) cold reactor (IGS, 2006) instead of a hot ITM (ion transport membrane) reactor, as in previous sections. The benefit of PSA technology with respect to ITM is the maturity. It has been widely used commercially for some decades. In short, the turbodiesel with a PSA oxygen reactor is titled as TD-2. Now, we wish to show the changes of the engine properties if it was to replace an ordinary diesel for the ZEMPES—TD-2 of the same capacity.

As a prototype for comparison, we selected the well-known good, high-speed diesel engine Ford FSD 425 with bore 93.67 mm, stroke 90.54 mm, volume 2.5 dm³, mean piston speed 12 m/s and effective power 52 kW for cars and lorries (Lurie, et al., 1985).

7.11.1   The Proposed Schematics

The proposed schematics of TD-2 are seen in Figure 7.12. They comprise a 2 cylinder H-shape engine and turbocompressor C1-T1 with heat exchanger H. In H, the

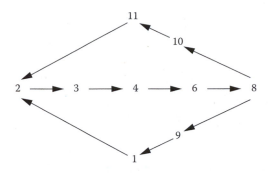

FIGURE 7.11 Flow directions.

waste heat is transferred from engine to incoming air. The unique feature of this system is the oxygen PSA reactor, denoted as AD.

It is commercially available from Innovative Gas Systems (IGS). Oxygen extraction is under pressure 3 to 4 bars.

The working substance of the thermodynamic cycle is an *artificial air*, the mixture of oxygen and carbon dioxide. It is permanently regenerated due to cooling combustion flue gases in R4, separation of water in WS, gas splitters GS1, GS2, GS3, combustion chamber VK, mixer Mi, and oxygen reactor AD.

Ambient air 18 enters compressor C1. After compression, the air is split into the two flows: one part 19 goes to H, the other 23 to AD. Pressures of both are equal, 3.5 bars. The processes in the main (closed) cycle are depicted as a rhomb (Figure 7.11).

The starting points are in flows 1 and 11. Processes 1–2 and 11–2 are going in both cylinders simultaneously as the open diesel cycles with equal flows and thermal energy are produced. The collector of flue gases AB is one for both cylinders, so processes 2–8 are in common: working substance is cooled due to giving heat to air.

In processes 8-9-1 and 8-10-11, the regeneration of the working substance takes place for each cylinder. In the auxiliary Brayton cycle on air, the ordinary fashion is 18-19-20-21-22.

The T-s diagram of both cycles (mainly on CO_2 and auxiliary on air) is given in Figure 7.13, where a-c is compression, C-Z' is isochoric heating, Z'-Z is isobaric heating, Z-b is combustion with expansion, b-2 is discharge.

The filling process 1-a and discharge b-2 are with some entropy gain due to the inherent irreversibility. Entropy decrease 2-3 is due to the cooling of the main working substance and the air heating. Decline of entropy 3-1 reflects the cooling and decline of flow rate.

In the auxiliary Brayton cycle, 18-19 is compression, then heating is 19-20, and cooling is 21-22 for recuperation of thermal energy.

Maximum cycle temperature in the cylinder of about 2150 K is normal for a diesel engine. It affects the walls only part-time. The steady-state temperature in point 12 after combustion chamber is 1642 K which is also admissible for an ordinary turbine steel.

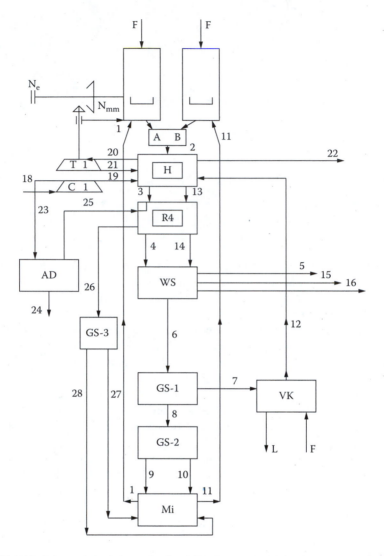

FIGURE 7.12 Schematics of the turbodiesel TD-2. F = fuel injection, N_e = engine shaft, N_{mm} = heat flow from losses in turbocompressor, AB = hot gas collector, H = heat exchanger, T1 = air turbine, C1 = air compressor, R4 = radiator, AD = oxygen PSA reactor, WS = water separator, GS–1, GS–2, GS–3 = gas flow splitters, Mi = mixer, VK = afterburner (combustion chamber).

These modest temperatures are much lower than in our turbodiesels previously described in Sections 7.7 and 7.8.

In TD-1 in an afterburner, all the content of oxygen in mixture (except some percent of residual oxygen) is used for combustion, whereas in proposed TD-2 only a fraction of available oxygen content is used. It means the possibility to demonstrate

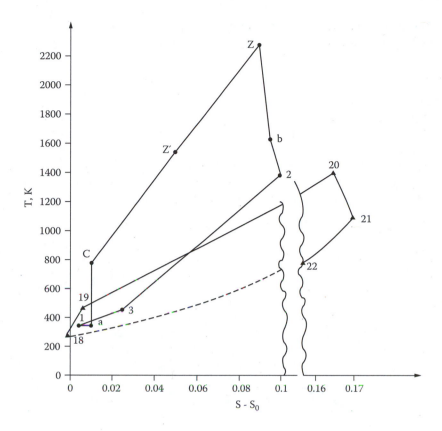

FIGURE 7.13 *T-s* diagram for the main (upper) and auxiliary (bottom) cycles. T in K, S in kW/K.

TD-2 in the near future is very conceivable as almost all ingredients are commercially available.

After the normal diesel operation process (displacement, suction, compression, combustion and expansion) flue gases CO_2, H_2O and excess of oxygen O_2 are going through collector AB. After cooling in H and R4 and dewatering in WS, the liquid water is deflected and the rest of the gases are split in GS1.

Flow 7 with some oxygen goes to afterburner VK to combust the fuel flow F with hot gases flow (12) to H. L is heat losses from afterburner surface cooling. The steam is condensed after cooling in H and R4, water is separated in WS, and deflected out of the cycle in 15. From WS is deflected (16) also that fraction of carbon dioxide, which formed in combustion is to be stored in a tank for subsequent sequestration. That way, no one combustion product or contaminant, dissolved in CO_2 is emitted into the atmosphere.

Flow 8 of CO_2 and O_2 is divided into two equal parts 9 and 10 in GS2, going to mixer Mi which is supplied with oxygen (28) and (27), and formed the artificial air flow 1 and 11 which entered the suction valves of the cylinders.

Oxygen is going (25, 26) from reactor AD to GS3 where it is split into two equal parts 27 and 28. Fuel injection in the cylinders is indicated as F.

7.11.2 Oxygen Separation from Air

The most important new feature of TD-2 with respect to TD-1 is the cold reactor AD based on pressure swing adsorption (PSA) technology.

Here are reproduced in Figure 7.14 and Figure 7.15 the technology and schematics of the PSA oxygen generator of the IGS Company (2006). Among many applications, there is no one related to zero emissions piston engine systems.

We have paid a great deal of attention to PSA oxygen generators as they are not well known in energy and automotive engineering. However, their use seems to be

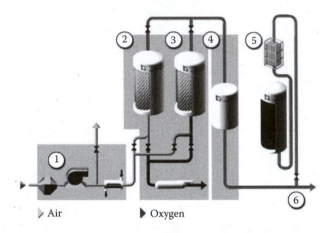

Air   Oxygen

The adsorptive separation of air is accomplished in the following process steps:
1. FEED AIR COMPRESSION AND CONDITIONING - The ambient air is compressed by an air compressor, subsequently dried by an air dryer and filtered before entering the process vessels.
2. ADSORPTION
The pre-treated air is passed into a vessel filled with Zeolite Molecular Sieve (ZMS) where most of the oxygen is passed through while nitrogen and other gases are adsorbed. Before the adsorption capacity of the ZMS is exhausted the adsorption process is interrupted.
3. DESORPTION
The saturated ZMS is regenerated (i.e. the adsorbed gases are released) by means of pressure reduction below that of the adsorption step. This is achieved by a simple pressure release system. The resultant waste stream is vented into atmosphere. The regenerated adsorbent is purged with oxygen and will now be used again for the generation of oxygen.
4. OXYGEN RECEIVER
Adsorption and desorption take place alternately at equal time intervals. This means that the continuous generation of oxygen can be achieved with two adsorbers, one being switched at adsorption and the other at regeneration. Constant product flow and purity is ensured by a connected oxygen receiver that stores the oxygen at purities up to 95% and pressures up to 4.5 bar(g).
5. OPTIONAL BACKUP SYSTEM
6. OXYGEN PRODUCT

FIGURE 7.14 Description of PSA technology by IGS Company (IGS, 2006).

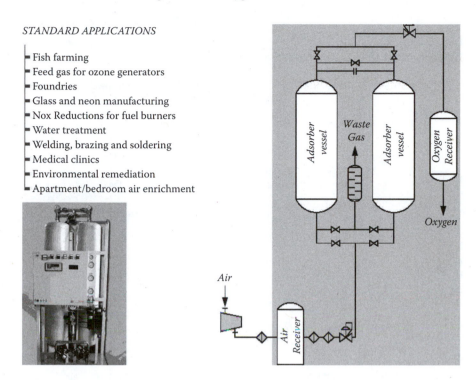

STANDARD APPLICATIONS

- Fish farming
- Feed gas for ozone generators
- Foundries
- Glass and neon manufacturing
- Nox Reductions for fuel burners
- Water treatment
- Welding, brazing and soldering
- Medical clinics
- Environmental remediation
- Apartment/bedroom air enrichment

FIGURE 7.15 Reproduction of a page of the IGS catalog on PSA oxygen generation (IGS, 2006).

very promising (in contrast to cryogenic oxygen generators) as they are easily transportable. Vibration and inertial forces are not detrimental to adsorption-desorption processes.

7.11.3 CALCULATION RESULTS

In Table 7.5, the parameters of the node points are given by oxygen excess coefficient 1.7 and oxygen fraction in O_2-CO_2 mixture of 0.4 and the pressure ratio 19.02 in each cylinder.

We maintain all of the cylinder equipment without any change with respect to prototype. Injection of high pressure fuel is assumed as the same, for example, by a plunger pump.

Table 7.10 demonstrates the oxygen flow rate 26 as 0.3787 mole/s or 12 g/s. In catalogues the capacity is given in normal m^3/hour, 12 g/s equals to 31.16 m^3/h of pure oxygen or 33 m^3/h of 95% O_2 by the air consumption about 360 m^3/h. It is well within the limits of all of the known oxygen producers, such as IGS Oxyswing: On Site Gas Systems, CAN Gas Systems, Grasys, Oxymat, to name a few. A steady-state demonstration of TD-2 might use a device from a catalog, such as '0100' or '0150' of IGS.

TABLE 7.10
Properties of Flows in the Node Points

Point	Pressure kPa	Temperature K	Flow rate mole/s	Enthalpy kW_{th}	Entropy kW_{th}/K	Substance
1	101.33	308	0.70033	7.967	0.002666	CO_2, O_2
2		1,344	1.5048	88.80	0.10120	CO_2, O_2, H_2O
3		455		27.25	0.02723	
5		308	0.2063	0.5441		water
6			1.29851	15.94	0.002673	CO_2, O_2
7			0.27656	3.394		CO_2, O_2
8			1.02196	12.54	0.002104	CO_2, O_2
9			0.51098	6.27		CO_2, O_2
10						
11			0.70073	7.967	0.002666	CO_2, O_2
12		1,643	0.29356	23.35		CO_2, H_2O
13		455	0.29356	5.647		
15		308	0.03402	0.0897		water
16			0.25955	3.411		CO_2
18		293	6.690	55.49	0.0000	air
19	354.64	440.4	4.878	61.51	0.007231	air
20	353.14	1,314		195.96	0.1739	air
21	102.00	1,011		147.46	0.1824	air
22	101.33	650		92.27	0.1152	air
23	354.64	440.4	1.81195	22.85		air
24	101.33		1.4333	18.07		nitrogen
25	354.64		0.37870	4.775		oxygen
26	101.33	308	0.37870	3.306		oxygen
27			0.18945	1.653		oxygen
28						

It is well known that almost all zero emissions power cycles have efficiency penalties of some percentage points with respect to their prototypes as the payment for cleanliness.

In our case, the power and efficiency are equal. The halved number of cylinders gives some space for a PSA oxygen generator. It might not be enough to setoff and extra volume would be the payment for cleanliness. Furthermore, extra weight and volume are needed for liquid CO_2 in a tank onboard, if TD-2 is used in transportation. Its mass is about 3 times more than the mass of fuel.

Calculations show the ability to create a zero emissions piston engine by converting an ordinary diesel into a turbodiesel TD-2 with PSA cold oxygen generator without sacrificing any power or efficiency. Additional mass and volume should be evaluated, which depend upon advances being made by oxygen reactor manufacturers.

TABLE 7.11
Comparison of the Prototype and Proposed TD-2

Parameter	Ford FSD425	TD-2
Coefficient of oxygen excess	1.7	1.7
Fraction of oxygen in mixture with inert gas	0.209	0.40
Compression ratio	19.02	19.02
Number of cylinders	4	2
Thermal power in cylinders, kW_{th}	147.41	2×67.35
Thermal power of afterburner, kW_{th}	0	20.08
Total thermal power, kW_{th}	147.41	154.78
Indicated power, kW	70.56	2×24
Indicated efficiency, %	47.87	35.63
Effective power of piston engine, kW	52.0	2×19.36
Turbine power, kW	0	48.5
Compressor consumption, kW	0	28.87
Mechanical losses in turbocompressor, kW	0	3.802
Added power from turbocompressor, kW	0	15.83
Effective power, kW	52.0	54.55
Effective efficiency, %	35.28	35.24
CO_2 production per hour, kg/h	39.41	41.11
CO_2 atmospheric emission, kg/h	39.41	0

7.12 TRIGENERATOR FOR ENHANCEMENT OF OIL RECOVERY (EOR)

The oil industry is the most attractive consumer of carbon dioxide. Currently, many millions of tons of crude oil are recovered in the U.S.A. by injection of carbon dioxide. Mainly, the source of CO_2 is natural, from the Earth's depths, such as in San Juan, Puerto Rico. However, a big project of the use of CO_2 from coal gasifier (Weyburn, U.S.A. Canada) is also in operation with success. For the schematics of EOR, see Figure 7.16.

Contemporary oil exploration needs much power for drilling, mud pumping, and oil lifting. Oil enhancing recovery uses not only water injection, but also carbon dioxide and steam. Elsewhere oxygen is used to inject into the depth of a well for combustion and oil viscosity reduction.

Generation of power and all mentioned agents at different devices is cumbersome and inefficient. The one system for all purposes seems to be much better. Earlier attempts to create such multigenerating systems are described as the OCDOPUS project by Yantovsky et al. (1993, 1994). Here is presented the last attempt, which is well known in the oil industry: diesel engines. First, the two principal elements are indicated; see Figure 7.17 and Figure 7.18. The total schematics can be seen in Figures 7.19 and 7.20.

Control valves I, II, III, IV for the three regimes of operation (A,B,C):

Some other valves might be used, distributing a flow from one channel into one of the two. Most attention needs the II valve after gas turbine T1 due to high temperature. It should be made of turbine blades-like steel.

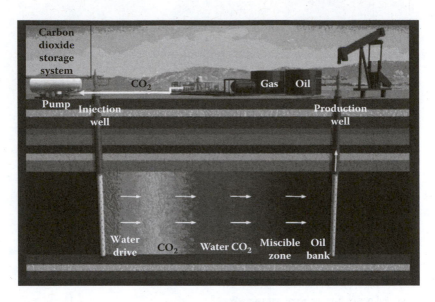

FIGURE 7.16 Enhancement of oil recovery (EOR) by CO_2 injection.

Valve positions and flow directions for the three operating regimes are:

A. Starting with emission to atmosphere. Engine shaft activated by an electrical generator, working as a motor from the grid.

   I—angular, 10"–10

   II— angular, 4– 4"

   III—angular, 5–5"

FIGURE 7.17 The cogeneration Wärtsila 12V34SG gas-fired piston engine power unit: electrical power 4040 kW, efficiency 46.1%, frequency 750 Hz, mass 76 t.

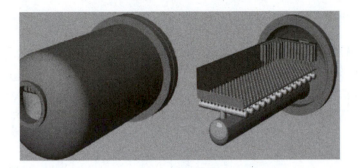

FIGURE 7.18 The unit for production of 350 t of pure oxygen per day (is to be halved for our case). The reactors are developed by Air Products and Chemicals (Armstrong et al., 2003), U.S.A.

B. Normal zero emissions operation with CO_2 and steam production.
   I—direct, 10'–10
   II—direct, 4'–4
   III—direct, 5–5'
   IV—direct, 29b–29
C. Power and oxygen production with atmospheric emissions.
   I—direct, 10'–10
   II—direct, 4'–4
   III—angular, 5'–5
   IV—angular, 29b–29'

Operation in B regime

Combustion products and CO_2 leave (3) cylinder of engine, expand in T1, and go for heating (4) the air flow (19). Air (18) is compressed in C2 and heated in H, then goes to MR, deposits oxygen (about 60%) and expands in T2. As it has a high enough temperature, it produces steam going to 23. Flue gases (4) are cooled first in H, then in R4 (flow 5). Being dewatered in WS, dry CO_2 enters splitter AB. A small part of the carbon dioxide, exactly equal to that born in combustion, is deflected (9) to be injected into a well. A major part of CO_2 is returned to Mi to be mixed with oxygen from (29), compressed in C1, cooled in R1, and enter the engine cylinder. Gas fuel flows from 12 and forms the combustible mixture in (1). Oxygen from MR (26) flows to R4 (28) and then to the mixer (29).

7.12.1 CALCULATIONS

Engine manufacturers do not indicate the important parameters of the engine. Here, they are calculated, using experimental data from a similar engine.

By supercharging, the pressure is 0.258 MPa, temperature working substance after cooling is 70°C and back-pressure after cylinder is 0.309 MPa, the volume efficiency is 0.844 and the cylinder charge (the sum of fuel, oxygen and inert gas) is equal to 181.89 mole/s.

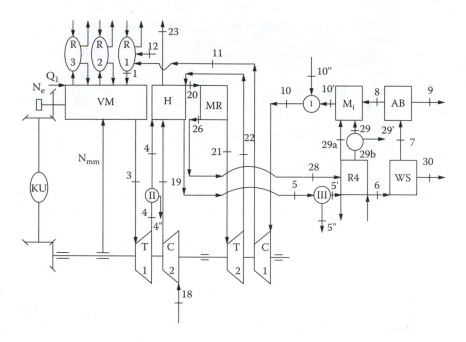

FIGURE 7.19 Schematics of the system. VM = piston engine, MR = membrane reactor, C1 = supercharging compressor, T1 = gas turbine, C2 = air compressor, T2 = depleted air turbine, H = air heater, WS = water separator, AB = gas splitter, Mi = mixer, R1-R4 = radiators (arrows are cooling air or water flows), N_{mm} = heat flow from turbocompressor mechanical losses, KU = clutch, N_e = mechanical power, Q1 = heat from fuel combustion, 12 = fuel gas inflow, 9 = carbon dioxide outflow, 30 = water outflow, 18 = ambient air inflow, 4" = starting flue gas outflow, 29' = oxygen production, 5" = flue gases outflow during oxygen production, 23 = hot air to steam boiler.

Assuming the fuel is natural gas with a heating value 827,317 kJ/mole, the calculated engine parameters are: supercharging pressure 0.258 MPa; maximal pressure 100 MPa; charge 181.89 mole/s; fuel flow rate 16.752 mole/s; oxygen flow rate 34.514 mole/s; inert gas flow rate 130.624 mole/s; thermal power 13,859 kW;

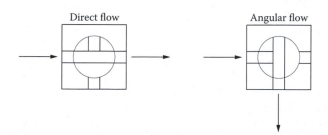

Direct flow   Angular flow

FIGURE 7.20 Valve position (the change of a regime needs to turn on 90 degrees).

TABLE 7.12
Node Point Parameters (see Figure 7.19)

Point	Pressure (kPa)	Temperature (K)	Flow (mole/s)	Enthalpy Flow (kW$_{th}$)
1	258.0	343	181.89	2,305
3	309.0	1,67	191.75	15,051
4	104.0	1,447	191.75	2,73
5	101.33	535	191.75	4,231
8	101.33	313	160.07	1,147
9	101.33	313	35.74	477.6
10	101.33	313	157.1	1,870
11	260.0	393.4	157.1	2,371
18	101.33	293	642.79	5,331
19	607.95	520.4	642.79	9.637
20	606.45	1,373	642.79	27,098
21	606.45	1,373	591.7	24,944
22	104.0	950.2	591.7	16,734.1
23	101.33	535.4	591.7	9,137.1
26	101.33	1,337	51.087	2,153.63
28	101.33	535.4	51.087	788.89
29	101.33	313	51.087	453.35
30	101.33	313	50.52	152.29

compression ratio 6.5; indicator power 4,724 kW; engine power 4,142 kW; effective power 5,315 kW; from turbocompressor power 1,173 kW; indicator efficiency 0.341; effective efficiency 0.384.

Effective power is the sum of the engine power and added from turbocompressor ones. To assume electrical generator efficiency of 0.97, the calculated electrical power is 4018 kW which exactly coincides with 4040 kW indicated by manufacturers in the catalog. This coincidence assures that other figures are calculated correctly.

In this project, the forcing of the engine by increase of the oxygen fraction in its mixture with CO_2 is recommended (Shokotov et al., 2006). In ambient air, the mass fraction of O_2 in mixture with nitrogen is 0.209. In this project, the recommended mixture ratio is $O_2/(CO_2 + O_2) = 0.3253$.

Calculated data on flows and parameters in node points of system are seen in Table 7.12.

In this regime, the system effective power is 9504 kW, efficiency is 46.3%, and carbon dioxide outflow is 5661 kg/h. The enthalpy flow (23) by air to steam boiler is 9137.1 kJ/s at 262°C heated water in boiler up to 222°C. By boiler efficiency 60% and steam enthalpy 2878 kJ/kg, the steam flow rate is 1.9 kg/s or 164 t/day.

The calculated power in kW is given next in Table 7.13.

TABLE 7.13
Calculation of Engine Performance

Indicator power of engine	5,189
Mechanical losses in engine	582
Gas turbine T1	2,32
Depleted air turbine T2	8,209
Supercharging compressor C1	501
Air compressor C2	4,305
Losses in turbocompressor	825
Effective power of system	9,504
Effective efficiency	0.463

TABLE 7.14
Summarized Results of the Analysis

Electrical power	9,219 kW
Electrical efficiency	44.9%
Carbon dioxide production	136 t/d
Oxygen production	139 t/d
Steam production at 1 MPa, 222°C	164 t/d

Notes: The simultaneous production of carbon dioxide and oxygen is impossible. Due to forcing, the generated power is doubled; hence, the electrical generator should be changed.

The substantial role of the depleted air turbine T2 is obvious.

By an electrical generator efficiency 0.97, the final results of the recommended system are given in Table 7.14.

REFERENCES

Armstrong, P., Sorensen, J., and T. Foster. 2003. "ITM Oxygen: An Enabler for IGCC," Progress Report: *Gasification technologies,* Oct. 12–15, San Francisco, CA, U.S.A.

Ashley, S. 2005. "On the road to fuel-cell cars," *Sci. Am.,* 3 (March), 50–57.

Ashley, S. 2007. Diesels come clean. *Sci. Am.,* 292(3) March: 62–69.

CeraMem Technology Brief, Dec. 2006. http://www.ceramem.com/ techbrief.pdf (accessed Oct. 2, 2007).

Doucet, G. 2007. "Electricity sector to play leading role in reducing CO_2 emissions," *Pennwell Global Power Review.*

Foy, K. and E. Yantovsky. 2006. History and State-of-the-Art of Fuel Fired Zero Emissions Power Cycles. *Int. Jour. of Thermodynamics* 9(2): 37–63.

Foy, K., Moran A., and J. McGovern. 2006. "A Mechanical Design for a ZEMPES Prototype," in *Proc. of Int. Conf. on Vehicles Alternative Fuel Systems and Environmental Protection* (VAFSEP), ed. Abdul Ghani Olabi, 56–60, Aug. 22–25, Dublin, Ireland.

Hoverton, M.T. 1964. *Engineering Thermodynamics*. New York: Van Nostrand, Princeton.

IGS Innovative Gas Systems. *Oxygen Generators*, http://www.igs-global.com/documents/ OxygenPSASystems_E1_001.pdf (accessed Oct. 7, 2006).

Jaichandar, S. and P. Tamilporai. 2003. "Low Heat Rejection Engines-An Overview," SAE Technical Paper Series 2003-01-0405.

Julsrud, S. et al., 2006. "Development of AZEP reactor components," in *Proc. of GHG-8 Conf.,* June 19–22, Trondheim, Norway.

Leidel, J.A. 1997. An Optimized Low Heat Rejection Engine for Automotive Use, *SAE Paper No. 970068.* http://personalwebs.oakland.edu/~leidel/ SAE_PAPER_970068.pdf (accessed Sept. 4, 2007).

Lurie, V. et al., 1985. "Car Engines." *Itogi Nauki I Techniki, Ser Dvig. Vnutr. Sgor.,* vol. 4, VINITI: Moscow, (in Russian).

McCormick, B. 2003. "The Hydrogen Energy Economy," presented at *Prepared Witness Testimony. The Committee on Energy and Commerce,* Subcommittee on Energy & Air Quality, May 20, http://energycommerce.house.gov/reparchives/108/ Hearings/ 05202003hearing926/McCormick1462.htm, (accessed Aug. 21, 2006).

McDonald, C.F. 2003. Recuperator considerations for future high efficiency microturbines. *Appl. Thermal Engng.* (23): 1463–1487.

Selimovich, F. 2005. "Modelling of Transport Phenomena in Monolithic Structure related to CO_2 - Emission Free Power," Lund Inst. of Technology, Sweden. http://130.235. 81.176/~ht/documents/lic-pres.pdf [accessed Oct 16, 2007].

Shokotov, M. and E. Yantovsky. 2006. "Forcing of Zero Emission Piston Engine by Oxygen Enrichment in Membrane Reactor," presented at *5th Annual Conf. on Carbon Capture and Sequestration.* DOE/NETL, May 8–11, Alexandria, VA, U.S.A., Paper 012.

Shokotov, M., Yantovsky, E. and V. Shokotov. 2005. "Exergy Conversion in the Thermochemical Recuperator of Piston Engine," in *Proc. Int. Conf. ECOS 2005*, June 20–22, Trondheim, Norway.

Shokotov, M., Yantovsky, E. and V. Shokotov. 2007. "Zero emissions turbodiesel with membrane reactor (ZEMPES-turbo project)," presented at *6th Annual Conference on Carbon Capture and Sequestration.* May 7–10, Pittsburgh, U.S.A. paper 063.

Sundkvist, S. et al., 2004. AZEP gas turbine combined cycle power plant, in *GHGT-7*, Univ. Regina, Sept. 5–9, Vancouver, Canada.

Sundkvist, S.G. et al., 2007. Development and testing of AZEP reactor components. *Int. Jour. of Greenhouse Gas Control* Vol. 1 (2), April: 180–187.

"Ultra Low Emission Vehicle" in Wikipedia – The Free Encyclopedia, [Online], http:// en.wikipedia.org/wiki/Ultra_Low_Emission_Vehicle (accessed Nov. 17, 2006).

Vente, J.F. et al., 2005. "On the full-scale module design of an air separation unit using mixed ionic electronic conducting membranes," Energy Res. Centre of Netherlands, Rep. ECN-RX-05-202. http://www.ecn.nl/docs/library/report/2005/ rx05202.pdf (accessed Oct. 9, 2006).

Weliang, Zhu et al., 2006. Mixed reforming of simulated gasoline. *Catalysis Today* (118): 39–43.

Yantovsky, E. and M. Shokotov. 2003. "ZEMPES (Zero Emission Piston Engine System)," presented at *2nd Annual Conf. on Carbon Dioxide Sequestration*, May 5–8, Alexandria, VA, U.S.A.

Yantovsky, E. and M. Shokotov. 2004. "ZEMPES (Zero Emission Membrane Piston Engine System)," presented at *2nd Annual Conf. On Carbon Dioxide Capture*, May 5–8, Alexandria, VA, U.S.A.

Yantovsky, E. and V. Kushnirov. 2000. "The Convergence of Oil and Power on Zero Emission Basis," *OILGAS European Magazine*, No. 2.

Yantovsky, E. et al., 1993. Oil Enhancement Carbon Dioxide Oxygen Power Universal Supply (OCDOPUS project). *En. Conv. Mgmt.,* 34(9-11): 1219–1227.

Yantovsky, E. et al. 1994. Exergonomics of an EOR (OCDOPUS) Project, *Enery-The Int. J.,* 12: 1275–1278.

Yantovsky, E. et al., 2005. "Elaboration of Zero Emission Membrane Piston Engine System (ZEMPES) for Propane Fuelling," presented at *Proc. of 4th Carbon Sequestration Conference.* May 5–8, Alexandria, VA, U.S.A., Paper 109.

Yantovsky, E. et al., 2005. Elaboration of Zero Emissions Membrane Piston Engine System (ZEMPES) for Propane Fuelling, in *4th Annual Carbon Sequestration Conf.* May 2–5, Alexandria, VA, U.S.A.

Yantovsky, E., Shokotov, M., McGovern, J. and V. Vaddella 2004. "A Zero Emission Membrane Piston Engine System (ZEMPES) for a bus," in *Proc. of Int. Conf. VAFSEP*, July 6–9, Dublin, 129–133.

8 Solar Energy Conversion through Photosynthesis and Zero Emissions Oxy-Fuel Combustion

8.1 BIOMASS COMBUSTION—IS IT A SUSTAINABLE ENERGY?

In looking for electricity generation in the future, we should keep in mind the fundamental restrictions seen today:

The possible eventual end of fossil fuels
The inability of the atmosphere to accept large amounts of combustion
    products
A strong opposition of people to any kind of nuclear power
A very small energy current density (large land demand) for solar and wind
    power
A deficiency of oxygen in the atmosphere

The natural question is where should energy come from? The only physical possibilities seem to be geothermal and solar energy. The first had its own restrictions but it is beyond the scope of this book.

In our view, the only source is solar as it does not consume fossil fuels and does not cause oxygen or nuclear reactions on Earth. The thermonuclear reactions on the Sun, with combustion of 4 million tons of hydrogen in helium per second are far away from us. This energy irradiates into space and a tiny part of it is falling on the Earth's surface supporting photosynthesis in plants on land and in the seas. A schematic representation of the balance between incoming solar energy and its exchanges between the Earth's surface and the atmosphere is shown in Figure 8.1.

A fraction of produced organic matter is called *biomass* and is used as fuel. With respect to this fuel, we should first consider the above question.

It is well known that exhaust gases of biomass combustion are no less harmful for our lungs (even carcinogenic) than exhausts of other hydrocarbon fuels. The only justification of biomass use is its CO_2 neutrality with respect to the greenhouse effect. All the CO_2 produced in combustion was previously absorbed in the photosynthesis reaction:

$$CO_2 + H_2O + light \Rightarrow CH_2O + O_2 \qquad (8.1)$$

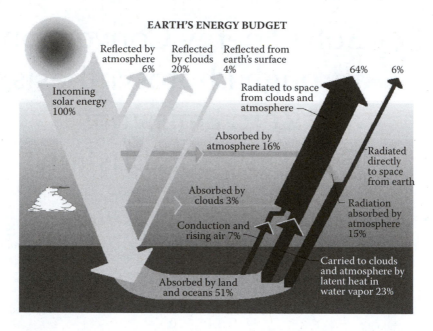

FIGURE 8.1 An illustration of the Earth's energy budget (value: 100% = 1368 W/m² is the solar constant) (from: http://marine.rutgers.edu/mrs/education/class/yuri/erb.html).

During combustion, we have:

$$CH_2O + O_2 \Rightarrow CO_2 + H_2O + heat \Rightarrow power \qquad (8.2)$$

Here, CH_2O is a simplified description of rather complicated organic matter in biomass fuel, like wood or peat pellets or green plants.

Along with other hydrocarbon fuels, such combustion of biomass violates a major human right—the right to breathe. Why should one breathe the dirty air polluted by biomass combustion?

That is why the answer to the title question is: evidently not. The improvement of this setting might be found if the same reactions (Equations 8.1 and 8.2) are used according to the zero emissions principle. This means the total elimination of discharge of combustion products into the atmosphere. Such a closed cycle of biomass use for electricity production was first described under the name SOFT cycle (solar oxygen fuel turbine) at the World Clean Energy Conference in Geneva, 1991; see Section 2.8. After detailed calculation, it was patented in the U.S.A., Russia, and Israel (U.S. Pat. 6,477,841 B1 of November 12, 2002). The schematic view of this SOFT cycle is shown in Figure 8.2.

The cycle is comprised of a pond with algae (the best are living in salty water macroalgae *Gracillaria* or *Ulva*), a fuel separating unit (FSU), an air separating unit (ASU) for oxygen production, a zero emissions power plant (ZEPP) and absorber of carbon dioxide, and some other combustion products by water from FSU, then returning them into the pond to feed the algae. This process is the subject of the present chapter.

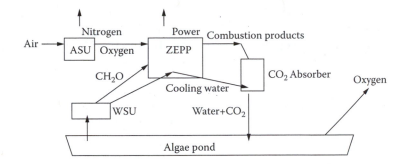

FIGURE 8.2 Schematics of the SOFT concept with an algae pond, fuel separation from water, air separation, oxy-fuel power plant and CO_2 absorption and return (Yantovsky, 2002).

The algae mass is growing, converting CO_2 into organic matter under solar light by photosynthesis. To increase profitability from algae, the high value organic matter as a raw material for the food industry, fodder etc. is to be extracted and the residual is used as the biomass fuel for power production. This is similar to the process in an oil refinery, where light fractions are first extracted. Due to elimination of nitrogen from the air, the products of combustion are easily absorbed by water and returned in the pond.

8.2 A SHORT HISTORY OF ALGAE CULTIVATION AND USE

Algae and seaweed use as a raw material is not new, especially in the food and pharmaceutical industries. Algae cultivation for electricity generation has been discussed in recent decades. Many countries are very interested now in the algae cultivation for liquid fuel production (see: http://www.oilgae.com/).

All algae have been divided by microalgae (size of some microns) and macroalgae or seaweeds, which are much greater. The photosynthesis is similar in both and our early history will start from microalgae. However, the technical problems of cultivation and combustion are different. That is why we will then focus on macroalgae only.

First published results of the use of open ponds with microalgae to convert carbon dioxide from power plants into methane fuel belong to Golueke and Oswald (Benemann, 1993). They demonstrated a small system, involving microalgae growth, digestion to methane and recycle of nutrients. They tried to catch CO_2 by injecting the flue gases into the pond regardless of a very small fraction of CO_2 in flue gases (about 10%). Especially active at that time was the Solar Energy Research Institute (SERI, now NERL) in the aquatic species program. After the testing of the three outdoor algae facilities in California, Hawaii, and New Mexico, it was concluded that it is possible to produce microalgae in a large-scale pond at high productivity and relatively low cost.

Similar results published by Alexejev et al. (1985) from Moscow University, demonstrate a small microalgae system called *Biosolar* with production of 40 g/sq.m dry biomass per day. The mineralized elements from the tank of produced methane are reused by algae. CO_2 is restored after burning. Alexejev stated: "1 mtce of methane might be produced from a surface of 70 sq. km annually."

Chemistry of the algae pond was described by L. Brown (1993, 1996) along with the outlook of a raceway-type pond and a paddlewheel to move water. He also stated:

> We estimate that microalgal biomass production can increase the productivity of desert land 160-fold (6 times that of a tropical rainforest). Microalgae require only 140–200 lb of water per pound of carbon fixed even in open ponds and this water can be low-quality, highly saline water.

If the pond water is rich with nutrients like a wasted municipal water or released from an animal farm, these very high figures of dry biomass production were published: 120 g/sq.m in a day (Lincoln, 1993), or 175 g/sq.m × day by Pulz (cited by Kurano et al., 1998). These figures translate into 40 to 50 kg/sq.m annually.

In parallel to the pond developments, some schemes of relevant power plants use of produced biomass as a fuel have been proposed. A patent by Yamada (1991) contains the use of dry algae as an addition to the regular fuel. A fraction of flue gases is released into the atmosphere by a stack. The rest is directed to an absorption tower to be washed by water, which dissolves CO_2 from the flue gases and returns it to the pond. A sticking point of this scheme is the rather small fraction of CO_2 in the flue gases, where the dominant gas is inert nitrogen. The separation of CO_2 from nitrogen turned out to be an insurmountable problem.

The radical solution is the separation of nitrogen before, not after, combustion which has been described by Yantovsky (1991), see Section 2.8. Further development of the SOFT cycle will be presented in the subsequent sections.

Combustion of biomass in the *artificial air*, the mixture of oxygen and steam or carbon dioxide, gives the flue gases without nitrogen. Carbon dioxide might be easily returned in the pond to feed algae.

8.3 WHAT IS ULVA?

Here are some reproduced data from the Galway Institute (Ireland): A green alga (up to 30 cm across) with a broad, crumpled frond is tough, translucent, and membranous; see Figure 8.3.

Crucial data for the SOFT project are productivity of ulva under natural insolation and by ordinary sea water temperature and chemical composition. There exists an experience of ulva harvesting in Irish land (Yantovsky and McGovern, 2006) where it is quite abundant. In addition to ulva, there exist a number of similar highly productive seaweeds.

Let us try to evaluate a possible growth rate of macroalgae with dimensions of a branch from one to ten millimeters. For simplicity, assume the form of organic matter particle as a sphere. Its volume is $V = (4/3) \pi \times r^3$ and crosscutting surface is $A = \pi \times r^2$. Solar energy flow density (insulation) is $\delta = 220$ W/m^2. Low heating value of produced organic matter is $H_v = 19$ MJ/kg, the biomass density $\rho = 800$ kg/m^3, efficiency of photosynthesis $\eta' = 10\%$. As the result of photosynthesis, the radius r is increased. According to standard definition of the relative growth rate RGR = M'/M,

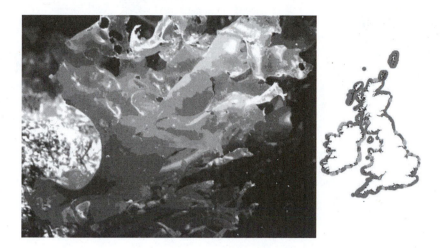

FIGURE 8.3 Ulva photo and distribution around Britain and Ireland (www.algaebase.org), (*Ulva lactuca*. Image width ca 6 cm (Image: Steve Trewhella. Image copyright information).

where M′ is the rate of mass increase in a second or in a day and M is mass of organic particle, we have:

$$\text{M′/M} = \text{RGR} = \frac{\delta_s \cdot A \cdot \eta'}{\rho \cdot V \cdot H_v} = \frac{3}{4} \frac{\delta_s \cdot \eta'}{H_v \cdot \rho \cdot r}, \tag{8.3}$$

In time increase $M(t) = M_0 \cdot \exp(t \cdot \text{RGR})$ (8.4)

Here it is assumed that RGR = Const.

In this formula, least known are the two quantities: the efficiency of photosynthesis (assume it as 10%) and the size of a considered particle (assume $r = 1$ mm). With these rather preliminary assumptions, we have

$$\text{M′/M} = \text{RGR} = (3/4) \times 220 \times 0.1/(19{,}000 \times 800{,}000 \times 0.001)$$

$$= 0.108 \times 10^{-5}\ [\text{s}^{-1}] = 0.00378\ \text{hr}^{-1} \sim 0.091/\text{day}.$$

The result is in agreement by order of magnitude with observed data. It is evident that the more r, the less is RGR. In some research, it is indicated that the decline of RGR after a large enough size of particles is achieved.

The direct measurement of the *Ulva lactuca* growth by different insolation in shallow water (40 to 70 cm) in the Roskilde Fjord, Denmark has been made by O. Geertz-Hansen and K. Sand-Jensen in 1992. The measured surface area A is initially 17 mm diameter ulva disks. Growth rates denoted μ_0 were calculated as:

$$\text{RGR} = \mu_0 = (1/t) \cdot \ln(A/A_o), \quad \text{where } t = \text{days of incubation} \tag{8.5}$$

Experiments vividly show the conversion of solar energy into chemical energy of the ulva biomass at the rather high latitude of Denmark; see Figure 8.6.

All five graphs are presented RGR in unit 1/day versus local insolation in mole/sq × m × day. The last unit should be converted in our convenient units W/sq × m. Here mole = mole of photons = 1 einstein = 210 kJ and day = 86,400 sec, hence, 10 mole/sqm × day = 24.3 W/m². Most important data are the rather high growth rate (up to 0.3 liters per day) in natural conditions of 55 grad of latitude by modest insulation and real temperatures.

Most productive seaweed ulva is already working for water cleaning (denitrification); see Table 8.1. New results on the maximum macroalgae growth shown in Table 8.2 were presented by Giusti and Marsili-Libelli (2005). The experience is of value for SOFT cycle.

TABLE 8.1
Ulva Production in the Denitrification Ponds

File Name: Ulva.xls Phytotreatment Pond Average Condition		File Name: Ulva.xls Phytotreatment Pond Average Condition	
Ulva biomass (kg fw m²)	1.5	Ulva biomass (kg fw m²)	1.5
Water depth (m)	1.0	Water depth (m)	1.0
Temperature range (°C)	15 ÷ 30	Temperature range (°C)	15 ÷ 30
Light intensity range ($\mu Em^{-2}s^{-1}$)	500 ÷ 2,000	Light intensity range ($\mu Em^{-2}s^{-1}$)	500 ÷ 2,000
pH range	7 ÷ 8	pH range	7 ÷ 8
Experimental Measurements		**Experimental Measurements**	
Ulva growth rate (d^{-1})	0.1	Ulva growth rate (d^{-1})	0.1
Ulva assimilation ratios (μmole N $d^{-1}m^{-2}$)	40,000	Ulva assimilation ratios (μmole N $d^{-1}m^{-2}$)	40,000
User-Friendly Tool		**User-Friendly Tool**	
INPUT	**INPUT**	**INPUT**	**INPUT**
Pond area (m²) enter value	1,300	Pond area (m²) enter value	8,000
Water flow (liter/s) enter value	250	Water flow (liter/s) enter value	140
Ammonia in inlet water (μM) enter value	179	Ammonia in inlet water (μM) enter value	61
Nitrate in pond water (μM) enter value	6.0	Nitrate in pond water (μM) enter value	6.0
OUTPUT		**OUTPUT**	
Biomass to be removed daily (kg fw d^{-1})	195	Biomass to be removed daily (kg fw d^{-1})	1,200
Denitrified nitrogen (%)	0.04	Denitrified nitrogen (%)	1.21
Assimilated nitrogen (%)	1.3	Assimilated nitrogen (%)	43.4
Total nitrogen removal (%)	1.4	Total nitrogen removal (%)	44.6
Ammonia in outlet water (μM)	1,76.5	Ammonia in outlet water (μM)	33.8

Source: Vezzulli, L. et al., 2006.

TABLE 8.2
Maximal Relative Growth Rate of Macroalgae and Rates of Other Processes

	Estimated Parameter Values and Confidence Intervals			
Parameter	Meaning	Unit	Literature Value	Calibrated Value
μ_{max04}	Ammonification rate	day^{-1}	0.045	0.1263–0.0251
μ_{max42}	Nitrification rate	day^{-1}	0.011	0.0010–0.00078
μ_{max23}	Nitrification rate	day^{-1}	0.046	0.1323–0.0147
μ_{denit}	Denitrification rate	day^{-1}	0.37	0.8329–0.0948
μ_{max}	Macroalgae maximum growth rate	day^{-1}	0.23	0.4509–0.0312
Ω_m	Macroalgae decay rate	day^{-1}	0.03	0.0594–0.062
SR	*Ruppia* decay rate	day^{-1}	0.041	0.0675–0.0043
ρ_{max}	*Ruppia* maximum growth rate	day^{-1}	0.17	0.3780–0.0235

Source: Giusti, E. and S. Marsili-Libelli, 2005.

As the depth of ponds here is 1 m, the dry weight of ulva biomass is 1.5 kg per cub.m of water and growth rate 0.1/day. Daily produced biomass is 1200 kg (case 'B') = 13.8 g/s. If one assumes the LHV of biomass = 19 kJ/g, the energy flow in biomass as a fuel is 262.2 kW. Assuming a realistic efficiency of fuel into power conversion as 25% (even in small units like microturbines or piston engines of ZEMPES), the produced power from such a pond of 0.8 ha surface is 65.5 kW or 100 kW from 1.22 ha. In the subsequent calculations, the same power needs 4 ha due to much less assumed biomass productivity. It is possible because the photosynthesis in denitrification is more productive than in sea water without nitrates.

A role of nitrides mentioned in earlier work (AFDW is an ash free dry weight, NGR = RGR) has been confirmed:

We recorded specific growth rates (NGR) ranging from 0.025 to 0.081 d^{-1} for a period up to two months in the repeated short-term experiments performed at relatively low initial algal densities (300–500 g AFDW m^{-3}). These NGR resulted significantly related to dissolved inorganic nitrogen (DIN) in the water column. Tissue concentrations of total nitrogen (TN) were almost constant, while extractable nitrate decreased in a similar manner to DIN in the water column. Total phosphorus showed considerable variation, probably linked to pulsed freshwater inflow.

In the long-term incubation experiment, NGR of ulva was inversely related to density. Internal concentrations of both total P and TN reached maximum values after one month; thereafter P concentration remained almost constant, while TN decreased below 2% w/w (by dry weight). The TN decrease was also accompanied by an abrupt decrease in nitrate tissue concentration. The biomass incubated over the two-month period suffered a progressive N limitation as shown by a decreasing NY ratio (49.4 to 14.6). The reciprocal control of ulva against biogeochemical environment and vice versa is a key factor in explaining both resource competition and successional stages in primary producer communities dominated by ulva. However, when the biomass

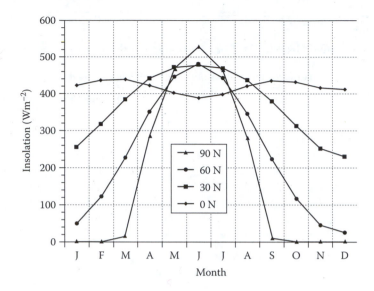

FIGURE 8.4 Averaged insolation on different latitudes: 0N = equator, 90N = North Pole.

exceeds a critical threshold level, approximately 1 kg AFDW m^{-3}, the macroalgal community switches from active production to rapid decomposition, probably as a result of selfshading, biomass density and development of anaerobic conditions within the macroalgal beds.

Systematic measurements of ulva growth in natural conditions of a coastal lagoon at Sacca di Goro, Adriatic Sea, Italy, have been made by Viaroli et al. (2005). On the area of 26 sq.km with an average depth of about 1.5 m by observing different chemical content of water, they recorded RGR of ulva about 0.05 to 0.15 1/day. This is a renewable source of fuel for the SOFT cycle of about megawatt range. Needed data on the monthly averaged insolation on different latitudes are given in Figure 8.4.

8.4 MACROALGAE AS A RENEWABLE FUEL

Having looked at the growth rate of RGR = 0.08–0.23 in literature and fantastic "calibrate value" RGR equal to 0.4509, we need to learn the main property of any fuel— the heating value. Sometimes, it is called *calorific value* when measured in calories. In literature, one may see different values ranging from 10 to 19 MJ/kg. What must be determined is what is meant by kg, dry or wet, and with ash or without. The most comprehensive assessment seems to be the work by M.D. Lamare and S.R. Wing (2001). Here, dry algae samples are disintegrated and combusted in a calorimetric bomb; see Figure 8.5.

Extrapolating to zero ash, we see 4.7 kcal/g dry weight or 19.67 kJ/g, which might be accepted for all organic matter of different algae. By 10% of ash, it is about

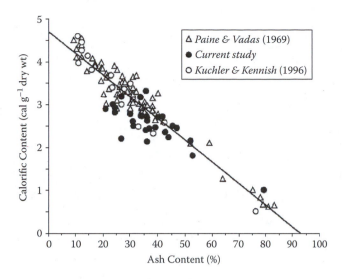

FIGURE 8.5 Correlation line for many algae: heating values versus ash content (Lamare and Wing, 2001).

19 kJ/g, which is selected for our energy conversion calculations. Heating value of algae depends on a season of growth; see Figure 8.6.

In these measurements, the heating value of ulva seems to be a little less than 19 MJ/kg. However, we will use just this figure as more statistically proven.

Quite recently, information came to light that Danish researchers look at seaweed for biofuel (see, http://biopact.com/2007/08/). In studies at the University of Aarhus, it was found that *Ulva lactuca* thrived when fed with liquid fertilizer and carbon dioxide, the greenhouse gas.

An optimal production process could yield up to 500 tons of biomass per hectare annually. The test species grew fast, doubling its biomass every three to four days, meaning RGR at about 0.25 of 1/day. It confirmed the measurements by Geertz-Hansen and Sand-Jensen (1992); see Figure 8.7.

In the SOFT process, the feeding by CO_2 and needed minerals such as returned ash are the unavoidable part of the total energy conversion.

Danish researchers say that for the time being, such high-tech forms of aquaculture as seaweed cultivation would be prohibitively expensive. We think that by being a part of the energy conversion process using large-scale equipment, the specific cost of biofuel could be reduced.

In the growing library of literature about biofuel production from plants or grasslands, the most popular is a giant grass miscanthus. According to data from the University of Illinois (http://miscanthus.uiuc.edu/index.php/research/), the annual yield of miscanthus in the U.S.A. is within the limits of 27 to 44 ton/ha. Having looked at this figure of annual ulva productivity of 500 ton/ha, we see a tenfold excess over the best crop on land due to much higher photosynthesis efficiency.

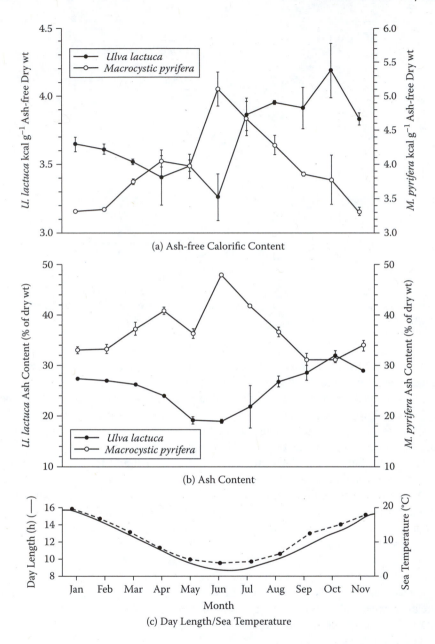

(a) Ash-free Calorific Content

(b) Ash Content

(c) Day Length/Sea Temperature

FIGURE 8.6 Heating value variation in a year versus a month of production (Lamare and Wing, 2001). (a) Ash-free, (b) ash content calorific content, and (c) day length in Otago Harbour, New Zealand. Note: The New Zealand winter season is May–August.

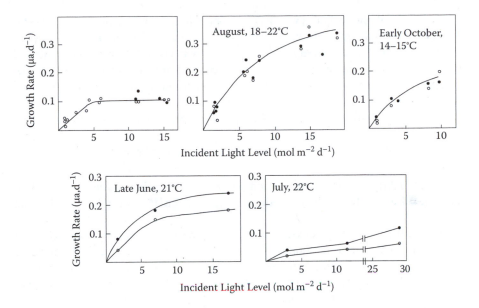

FIGURE 8.7 Ulva growth rate versus insolation near Denmark seasonal variation (Geertz-Hansen and Sand-Jensen, 1992).

8.5 MACROALGAE CULTIVATION IN ISRAEL AND ITALY

The crucial data are based on Israeli experience (Osri, 1998). In 1998, there were three raceway-type ponds, each having a surface of 1500 sq.m with paddlewheel sea water circulation. CO_2 was supplied by a tank on a lorry and injected into the water by perforated tubes. The depth of the water was 0.4 m, having a power hydrogen factor pH = 7. The firm figures were obtained for seaweed gracilaria only. The stable productivity of dry mass from a pond was 12 t/year or 8 kg/sq.m/year. These ponds were located in Northern Israel, near the sea, from which the sea water was pumped into ponds. The produced biomass was used as raw material for chemicals and pharmaceutics. Recently, some headway in seaweed cultivation has been made by Noritech-Seaweed Biotechnologies Ltd (http://www.noritech.co.il/Noritech/index.asp).

In Italy, the main practical interest in ulva seems to be concentrated in water cleaning and denitrification (De Casabianca et al., 2002; Bartoli et al., 2005; Vezzulli et al., 2005) where much research has been done in Genova, Venice, and Parma universities. Their active work gives an opportunity to use the SOFT cycle as an incinerator, deflecting extra nitrides, heavy metals, and other contaminants and also as a fuel separation device to dispose of contaminants, perhaps in some depth underground.

8.6 ENERGY FLOW CONCENTRATION

The main obstacle of solar energy capture is its very low current density, especially when annually averaged. In Israel, it is about 220 W/m², only 16% of the solar constant (1368 W/m²). In central Europe, it is half that much. Thus, the energy expenditure and cost of incidental energy absorption is of primary importance.

In the case of photovoltaics with an algae pond, it is much less, about 3 to 5%. But the energy expenditure of absorber is: pond is 100 times less than that of silicon.

Having been absorbed by algae, the solar energy in chemical form is concentrated by water flow much better than by an optical concentrator. The concentration factor of a paraboloid concave mirror is about 500; it means the averaged focal spot energy current density is about $500 \times 220 = 110 \text{ kW/m}^2$.

Energy flow in the pipe from the algae pond to processing is about:

$$\alpha \times \rho \times w_w \times H_v = 0.001 \times 1{,}000 \times 1 \times 19.106 = 19{,}000 \text{ kW/m}^2 \qquad (8.6)$$

Here, $\alpha = 0.001 =$ mass fraction of biomass in water, $\rho = 1000 \text{ kg/m}^3$, in water density, $w_w = 1 \text{ m/s}$ — flow velocity of the water, $H_v = 19 \text{ MJ/kg} =$ dry biomass heating value.

It is evident that energy current density in the pipe is one hundred times more than that in the focal point of an optical concentrator. It is referred to as hydrodynamic concentration. It means the equipment size for the subsequent energy conversion processes should be rather small. It is more important than the large size of a solar energy absorber (inexpensive pond or shallow sea under offshore wind turbines).

Let us evaluate what height h and what velocity w_w might achieve in one ton of water using seaweed contained in it for energy. Assume the efficiency of separation fuel from water as 0.9, and efficiencies of power cycle as 0.5 and that of hydraulic pump as 0.8, by heating value of dry mass $H_v = 19 \text{ kJ/g}$ and its concentration 0.001. From simple relation:

$$M \cdot g \cdot h = \eta_m \cdot Hv, \qquad (8.7)$$

we have $H = 0.9 \times 0.5 \times 0.8 \times 0.001 \times 19{,}000{,}000/9.81 = 697 \text{ m}$.
Based on the kinetic energy $M \times w_w^2/2$, it follows

$$M \cdot w_w^2/2 = \eta_m \cdot H_v, \qquad (8.8)$$

and

$$w_w = \sqrt{0.36 \cdot 0.001 \cdot 19{,}000{,}000} = 82.7 \text{ m/s}$$

These figures show that even a small part of energy in water with biomass is enough to elevate this water to the height of the highest dams or to create its flow with rather high velocity. It is important to regulate the appropriate circulation of water in the pond to give time for seaweed growth.

8.7 POWER UNIT OUTLOOK

A schematic of the SOFT cycle is presented in Figure 8.8. Water with algae 6 from the pond 4 is going to the water separation unit 12, from where the pure water without algae is used as a circulating water to cool condenser 14 and absorb CO_2 in 16.

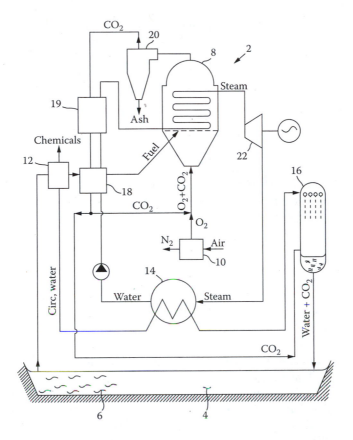

FIGURE 8.8 Schematics of the SOFT cycle (Yantovsky, 2002).

Wet organic matter is dried in 18 by heat of flue gases. Relatively dry fuel is directed to the fluidized bed combustor 8. After combustion in the artificial air (the mixture of oxygen and carbon dioxide) flue gases go in the cyclone separator 20, the deflected ash is returned into the pond CO_2 with some steam going through the heat exchanger 19 and fuel drier 18 to a separation point, from where a major part is mixed with oxygen, forming artificial air for fluidizer and a minor part is directed to absorber 16 to be dissolved in circulation water and returned to the pond. This minor fraction of CO_2 flow is exactly equal to CO_2 appeared in combustion. Oxygen is produced from air at the cryogenic or ion transport membrane (ITM) unit 10. Water from condenser 14 goes by a feed water pump through heat exchangers 18 and 19 into tubes of the fluidized bed combustor 8 (boiler). Produced steam expands in the turbine 22, driving the generator. Low pressure steam is condensed in 14. Actually, it is the ordinary Rankine cycle.

Some word on the chemical production. It is unwise to combust the crude seaweed at a power plant in the same sense as it is equally unwise for such use of crude oil. A small mass fraction of seaweed contains very useful organic chemicals which should be deflected along with water separation before the fuel combustion. There exist lots of methods for high organic separation, which is far from the scope of the present book. In any case, the chemical production could improve the economics of the SOFT cycle.

Let us take for a numeric example, the decentralized power supply by a small power plant of 100 kW (Iantovski et al., 1997). In order to get reliable figures, we make rather modest assumptions:

Fuel is wet (50% water content)
ASU power consumption by 98% oxygen purity 0.22 kWh/kgO$_2$
Superheated steam before turbine 130 bars, 540°C,
Isentropic coefficients of turbine 0.80, of feed pump 0.75
Seaweed productivity 16 kg/sq.m/year or 10 W(th)/sq.m
Photosynthesis efficiency 4.6%

Calculated results are:

Heat input	425.5 kW$_{th}$
Net output	107.3 kW$_e$
Cycle efficiency	25.2%
Pond surface	4 hectare

The graphs of efficiency versus fuel moisture are outlined in Figure 8.9. For possible figures of Rankine cycle with reheat and efficiency of 35%, the needed surface of the pond is 3 ha. To supply an Israeli kibbutz of some hundreds of people, 4 to 5 such units and a pond of 15 to 20 ha is sufficient.

For a local power plant of 10 MW with the cycle efficiency of 40% and photosynthesis efficiency of 6%, the specific power per square meter is about 5 W

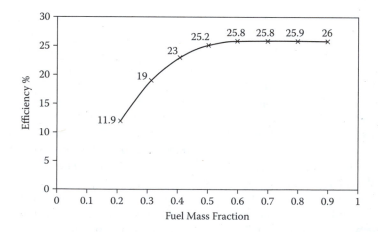

FIGURE 8.9 Efficiency of a 100 kW$_e$ cycle vs. dry fuel mass fraction in a wet fuel (Iantovski et al., 1997).

$(220 \times 0.4 \times 0.06 = 5.28)$, and the pond size is about 2 sq.km. By order of magnitude, it is comparable to the Yatir reservoir in the Negev Desert near Beer Sheva. The Keren Kayemeth LeIsrael is planning to build 100 water reservoirs in the next 5 years. One of these might be used for the SOFT demonstration.

Finally, for the national power demand of 10 GW (about 2 kW per capita) in Israel, a reasonable extrapolation is possible: expecting specific power of 10 W/sq.m due to the increase of cycle efficiency and photosynthesis. It means the needed pond surface is about 1000 sq.km. The surface of the Dead Sea is identical (exactly 980 sq.km). If in the future a Life Sea (with the normal, not deadly salt concentration for seaweed) would appear in the desert, not too far from the Dead one (see Figure 8.10), it could give the country full electrical power along with a lot of fresh water and organic chemicals. There would be no emissions of combustion flue gases and no net consumption of oxygen, which is consumed in combustion but released in photosynthesis. The only need is solar energy and a piece of desert. With either, a possible terror attack would not cause any serious damage.

An Israeli representative at the Johannesburg Summit, Jacob Keidar, announced the Israel-Jordan project of a 300-km long pipeline t from Red to Dead Seas. The Life Sea might be a useful consumer of the transferred water at the middle of the pipeline; see Figure 8.10.

8.8 GASIFICATION

In the proper energy mix, not only electricity, but also gaseous or liquid fuel is needed. In the SOFT cycle, it is attainable by a small modification (Figure 8.11). The difference is the incomplete combustion (gasification) in the fluidized bed reactor. Now it is a gasifier 24. Biomass gasification is well documented.

Fluidized bed gasification experiments with the sugarcane bagassa described by Gomez et al. (1999) produced a gaseous fuel mixture consisting of carbon monoxide, hydrogen, and carbon dioxide. After cleaning in 20, it is used in a piston engine or turbine 26, producing mechanical power. The same fuel–gas mixture might be converted into liquid fuel like methanol or even gasoline. After combustion in 26, the flue gases are absorbed by circulated water and returned to the pond 4 to feed seaweed 6. The figures in brackets 0.06 and 10^3 reflect mass ratio water/CO_2.

Similar processing of seaweed biomass is possible by thermolysis, when heating it in a closed volume by flue gases.

8.9 WATER DESALINATION

For the state of Israel, the problem of fresh water is no less severe than that of electrical power supply. The annual demand is about 1.4 cub.km of fresh water. It rains only 50 days a year and 60% of the land is deserts.

Let us consider what the SOFT cycle might do for desalination: is it possible to use low-grade heat after the turbine expansion to evaporate a fraction of the circulating salty water (sea water) with the subsequent condensation of vapor for the fresh water production (desalination)?

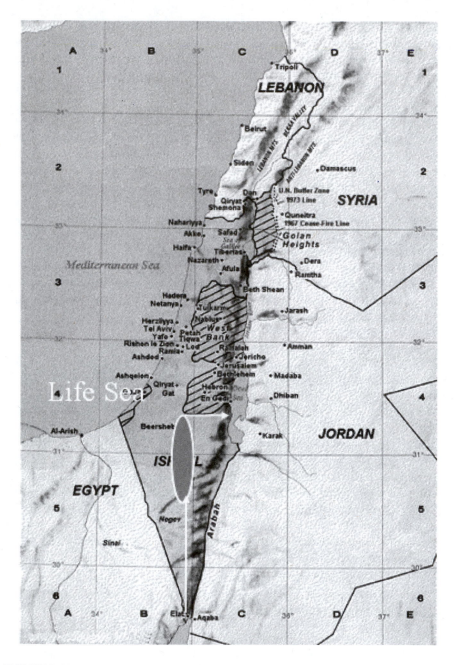

FIGURE 8.10 Location of the tentative Life Sea in the Negev Desert in the middle of the Red Sea — Dead Sea Channel.

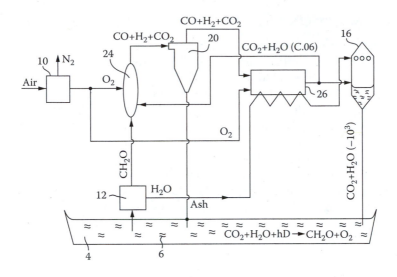

FIGURE 8.11 Gasification version of the SOFT cycle (Yantovsky, 2002).

Assume an evaporator of a minor fraction of circulating water after turbine. Cooling and condensing this vapor by the major part of circulating water gives fresh water as condensate.

How large is its flow rate? Assume the turbine as a back-pressure type, by exit steam pressure 1.2 bars. If in a modern high-temperature steam turbine the inlet is 1000 K by 200 bars, the enthalpy is 3874 kJ/kg. After expansion, the steam is at 450 K and 2830 kJ/kg. For water evaporation by 1 bar, the enthalpy drop of 2500 kJ/kg is enough.

In a small power unit of 100 kW, the mass flow rate of cycle water of a Rankine cycle is $100/(0.25 \times 1044) = 0.4$ kg/s. The mass flow rate of desalinated water is the same 0.4 kg/s. For a small demonstration plant, the figures are: pond surface 4 ha (40,000 sq.m); power 100 kW; dry fuel flow 0.021 kg/s; chemicals (4%) 1 g/s, and the fresh water 0.4 kg/s.

Specific dry fuel consumption is 756 g/kWh. It is about twice the excess of standard fuel consumption in microturbine power units due to lower heating value and low efficiency.

In a 1 GW power plant with cycle efficiency 40% and pond surface 10×20 km, the flow rate of produced fresh water is 4 Mg/s or 14,400 t/h. Assuming operation at 7,000 h/year, the yield of water annually is about 0.1 cub.km. It is evident that if the SOFT cycle with water desalination would be used on a full scale, it might meet all water demand.

Contemporary practice of the use of 18 power generating and desalinating plants at the West Bank of Arabian Gulf Asouri, giving 15 GW of power and 1.9 cub.km of desalinated water annually, confirms the above assumptions.

In case of applicability, the experimental results of Italian researchers with higher growth Sulva figures, the size of the mentioned ponds might be greatly reduced.

8.10   COMPARISON WITH THE FIRST SOFT VERSION OF 1991

The closed cycle power plant concept, based on algae photosynthesis in a pond, combustion of organic matter of dried algae in a zero emissions power plant and CO_2 capture to return in the pond for feeding algae was published in 1991; see Figure 10 in Section 2.8 (Yantovsky, 1991). Here, an air separation and expansion in a steam turbine was used. The difference was in the inert gas, which replaced nitrogen in the combustor. It was not carbon dioxide but steam. Also different was the algae: micro, not macro, was used. A fluidized bed combustor was not used but was replaced by a gas-turbine one for clean fuel. After the triple expansion in turbines together with steam, the carbon dioxide was returned to the pond.

Now this version is actively used by Clean Energy Systems (CES) creating a demonstration plant of 5 MW in California not for algae, but for ordinary gas fuel. It is the first zero emissions power plant (Figure 2.5). Had it been successful, it might be added by an algae fuel system for a SOFT cycle demonstration.

It is confirmed that seaweed ulva, selected as a renewable fuel for the SOFT cycle is well documented, its main properties are: relative growth rate RGR = 0.1 to 0.2 per day (or 10–20 times in a month by averaged insolation) and heating value of 15 to 19 MJ/kg of ash-free dry weight. Optimal organic matter concentration is 1kg/cub.m (0.001 by mass of water).

The SOFT power cycle is of practical interest to countries with sufficient solar radiation and this concept is ready for engineering, economic calculations, and demonstration.

It is non-fossil fuel, non-nuclear, non-polluting, and non-oxygen consuming power cycle with the least expensive receiver of solar radiation and effective hydrodynamic concentration of energy flow. Its additional service to the human environment might be incineration by combusting ulva with nitrides and other contaminants from added brackish water.

REFERENCES

Alexejev, V.V. et al., 1985. "Biomass of microalgae use for solar energy conversion," in *Techno-Economic and Ecology Aspects of Ocean Energy Use*. TOI Vladivostok, Russia, 53–58 (in Russian).

Azoury, P.H. 2001. Power and desalination in the Arabian Gulf region: an overview, *Proc. Instn. Mech. Engrs.* Part A, 215 (4): 405–419.

Bartoli, M. et al., 2005. Dissolved oxygen and nutrient budgets in a phytotreatment colonised by *Ulva* spp, *Hydrobiologia* 550(1): 199–209.

Benemann, J. 1993. Utilization of carbon dioxide from fossil fuel burning power plants with biological systems. *Energy Conv. Mgmt.* 34(9-10): 999–1009.

Brown, L. and K. Zeiler. 1993. Aquatic biomass and carbon dioxide trapping. *Energy Conv. Mgmt.* 34 (9-10): 1005–1013.

Brown, L. 1996. Uptake of carbon dioxide from flue gas by microalgae. *Energy Conv. Mgmt.* 37(6-8): 1363–1367.

De Casabianca, M.L. et al., 2002. Growth rate of Ulva rigida in different Mediterranean eutrophicated sites, *Bioresource Technology* 82 (1): 27–31.

Geertz-Hansen, O. and K. Sand-Jensen. 1992. Growth rate and photon yield of growth in natural populations of a marine macroalga Ulva lactuca. *Marine Ecol. Progr. Ser.* (81): 179–183.

Giusti, E. and S. Marsili-Libelli. 2005. Modelling the interaction between nutrients and the submersed vegetation in Ortebello Lagoon, *Ecological Modelling* 184(1):141–161.

Gomez, E.O. et al., 1999. Preliminary tests with a sugarcane bagasse fuelled fluidized bed air gasifier. *Energy Conv. Mgmt.* 40(2): 205–214.

Iantovski, E., Mathieu, Ph., and R. Nihart. 1997. "Biomass fuelled CO_2 cycle with zero emission," in *Proc. of Powergen Europe*, June 17–19, Madrid, Spain.

Kurano, N. et al., 1998. Carbon dioxide and microalgae, in: T. Inui, et al., (eds.): *Advances in chemical conversions. Studies in Surface Sci. and Catalysis.* Elsevier, (114): 55–59.

Lamare, M.D. and S.R. Wing. 2001. Calorific content of New Zealand marine macrophytes, *New Zealand Jour. of Marine and Freshwater Res.* (35): 335–341.

Lincoln, E. 1993. *Bulletin de l'Institut océanographique*, Monaco, No.12, 141–155.

Olsson, et al., 2000. "Cogeneration based on gasified biomass," in *Proc. of Conf. ECOS 2000*, University Twente, July 5–7, Enschede, the Netherlands, vol. 4, 1945–1957.

Osri, U. 1998. "Seaweed cultivation project in Israel. Rosh Hanikra," (private communication).

Vezzulli, L. et al., 2006. A simple tool to help decision making in infrastructure planning and management of phytotreatment ponds for the treatment of nitrogen-rich water. *Water SA,* vol. 32, No. 4, 605–609.

Viaroli, P. et al., 2005. Nutrient and iron limitation to Ulva blooms in a eutrophic coastal lagoon. *Hydrobiologia* (550): 57–71.

Yamada, M. 1991. "Recovery and fixation of carbon dioxide," Patent of Japan 03154616. Appl. Nov. 10, 1989, publ. July 2, 1991.

Yantovsky, E. 1991. "The thermodynamics of fuel-fired power plants without exhaust gases," in *Proc. of World Clean Energy Conference CMDC*, Nov. 4–7, Geneva, Switzerland, 571–595.

Yantovsky, E. 2002. "Closed Cycle Power Plant," U.S. Patent No. 6,477,841 B1, Nov. 12, 2002.

Yantovsky, E. and J. McGovern. 2006. "Solar Energy Conversion through Seaweed Photosynthesis with combustion in Zero Emission Power Plant," in *Proc. Conf. Renewable Energy in Maritime Island Climates*, Apr. 26–28, Dublin, Ireland, 23–27.

9 Associated Tool for Calculations

INTRODUCTION

We have considered different cycles, each with different working substances and different parameters. What should a designer do? How to select the best solution?

Among many goals and targets, we will fix the three values to be minimized: fuel consumption, money expenditure, and pollution. As usual, when one of the three is decreased, the others might increase. For decision making in modern energy policy, the optimization is multicriterial. Many decisions change criterions in opposite directions.

At the end of this chapter is a presentation of the method of optimization of Wilfredo Pareto: multicriterion optimization in the frame of three axes: fuel, cost, and pollution. The cost is the subject of economic calculation on the lowest of current and forecasted prices. It is definitely beyond the scope of this book. Pollution is the subject of chemical engineering, treating energy equipment as chemical reactors. Fuel would only be considered in some detail as it is considered thermodynamics. Instead of fuel minimization, we will consider a more general approach, well developed in modern engineering thermodynamics—the exergy analysis. The next section is written to be more understandable for newcomers to exergy.

9.1 WHAT IS EXERGY?[*]

> —*What is Matter?*
> —*Never mind.*
> —*But what is Mind?*
> —*It doesn't matter.*

> **From a talk between a curious boy and his grandfather**

When we went to school, rather long ago, we learned physics lessons about mass, force, work, energy, charge, and other important things. But I had never heard the word "exergy."

Now, when we are writing these lines, we keep in mind our children and grandchildren. We wish to do our best to tell them and others, and, perhaps, even their teachers, what exergy is, in order to include it in the list of the above mentioned terms.

To tell fundamental physical terms in the form of their definition, using other, more understandable words, is almost impossible. There are no words more simple than mass or energy. Examples, taken from everyday experience or seen in nature or engineering might help to understand "what it is." The term "exergy" should be

---

[*] Reproduced from the book "*Energy-Efficient, Cost-Effective, and Environmentally-Sustainable Systems and Processes*" Edited by R. Rivero et al. pp 801–817, Instituto Mexicano del Petroleo. Printed in Mexico. Copyright 2004.

introduced to teenagers, regardless of their future profession. Otherwise, their education would be incomplete. Exergy will help to form the *weltanschaung* of young people toward sustainable world development.

9.1.1 Natural Questions

Something is rotten in the state of Denmark.

Shakespeare, *Hamlet*

Something is really not in order (rotten) in the fundamental definitions. One may find such a wording in many textbooks for schoolchildren: *Energy is the ability to do work.* If one is so curious as to search what work is, he finds the answer: *Work is energy in transition …* or about the same, … *energy transfer across a boundary.* Such definitions form a vicious circle. The history of physics shows vicious circles occur to cover misunderstandings.

Other questions appear when a pupil learns energy conservation law from a teacher. "I have heard many times we are obliged to save energy, to prevent its wasting because it is very difficult to produce it. Why do I need to conserve energy if our Lord actually does this? Would He like to do the same in the future as He did it in the past? And how is it possible to produce energy if it can't appear or disappear? I feel I need to conserve something, but what it is?"

These questions are not naïve. However, they are missing in secondary school physics. Here, I'm trying to fill the gap. All the things I wish to discuss are considered using examples from everyday life, almost without any mathematics. I think the best way to explain a general law is to recognize a similarity or coincidence in some examples which at first glance appear to be quite different. The first example we will use to show the main assumption under consideration is the assumed equilibrium of the state of considered events. The difference between it and nonequilibrium is evident in Figure 9.1.

9.1.2 Mountain Bike

Great things are done when men and mountains meet.
This is not done by jostling in the street.

William Blake

You are standing with your bike at the top of a hill. When going down, even without using pedals, you will move with significant velocity. Your mass (you and the bike) times half of the square of velocity nearly equals your weight times the height difference between the top of the hill and your actual position at the moment. Why do we need to underline "nearly equals"? It is to denote exactly equal if there is no friction.

This example of friction shows us an everyday experience: there always exists a force directed opposite to velocity. In the case of a bike, it is the friction of the wheels

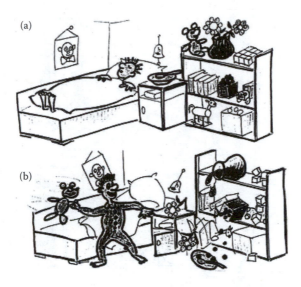

FIGURE 9.1 Equilibrium (a) and nonequilibrium (b) state in a very familiar thermodynamic system.

and resistive action of air. Due to this friction that we will call "mechanical," you can never reach the same height on the next hill without using pedals.

We used to say that on the top of a hill, you have a potential energy. Going down, you have a kinetic energy, which nearly equals the change of a potential one. We expect the conservation of energy according to the law of its conservation, as stated by Robert Julius Mayer in 1842. The deficiency equals the friction work, which is friction force times your path downward. The natural question of a curious boy is, "where is the deficiency of kinetic energy going?" If you rub your palm over a carpet, you feel the temperature rise: friction produces heat. The heat might be very intensive. It might melt metals in friction welding devices. In Figure 9.2a, friction is the total work done by a man rotating the pedals. His eager wife is using friction to boil some water for cooking.

The first experiments showing the equivalence of friction work and heat were made by James Prescott Joule. In his lecture "On Matter, Living Force and Heat" in St. Anna Church Reading Room, Manchester, on April 28, 1847, he stated:

> Experiment had shown that wherever living force is apparently destroyed or absorbed, heat is produced. The most frequent way in which living force is thus converted into heat is by means of friction. In these conversions nothing is ever lost. The same quantity of heat will always be converted into the same quantity of living force.

Here, living force is what we now call kinetic energy.

We wish to add only one comment to these very important words, found in A. Lightman's excellent book, *Great Ideas in Physics* (Lightman, 1992). The last

FIGURE 9.2 Work conversion into heat by friction.

sentence clearly expressed the energy conservation law, but contradicts another, not less important law of nature, the law of energy degradation. Actually, all heat can never be converted into kinetic energy of moving bodies, which was discussed by J. Joule. Only a portion of heat can be converted. We will speak about it a little later.

We need to accept that mechanical friction produces some heat at the temperature of sliding surfaces.

9.1.3   WATERFALL

> *... and again I hear*
> *These waters, rolling from their mountain spring*
> *With a soft inland murmur.*

W. Wordsworth Lines Composed a Few Miles above Tintern Abbey

We see more or less the same phenomenon looking at a waterfall. The water falling from a dam, which was at rest before the dam, might flow with a great velocity. Its kinetic energy *nearly* equals to potential one before the dam. The difference is caused by so-called "hydraulic friction," the shear of water layers near the channel wall. Behind the dam, at the bottom, where water is again at rest, its temperature should be a little higher; however, in reality, its rise is very small. The same great man, James Joule, during his vacation in Switzerland, tried to measure it with a thermometer with 0.1°C scale and failed to detect any rise.

In order not to lose the energy of water, but instead to use it to produce power, a turbine should be installed in the falling water. Turbine blades rotate and convert the potential energy of water into shaft work. Then in an electric generator, some power is produced. The obtained electrical energy is less than the energy of falling water due to the two kinds of friction: mechanical friction of water over the walls, turbine blades, and lubricants in journal bearings and a new important kind of friction, electrical friction in conductors.

We can transfer electrical charges in solid (metals), liquid (solutions), and gaseous conductors. The flow of electrical charges we call current carries electrical power. In most cases, the amount of positive and negative charges is strictly equal. It means the conductor as a whole is electrically neutral. In metals, the positive charges are fixed in the crystalline lattice, wherever the negative ones (electrons) form electron "gas," which flows in a conductor similar to a neutral gas in a pipe, with the lattice as an obstacle. Electrons collide with fixed particles, which create a kind of retarding force for electron gas. We call it *electrical friction*. It is actually the same as electrical resistance, which produces Joule's heating of a conductor. We use this heat in numerous cases at home, i.e., when ironing trousers or boiling water. The last is depicted in Figure 9.2(b) when electricity is produced by a man, performing physical exercises. This heating might increase the temperature of a refractory metal up to 3000 K, as can be seen in a filament of a light bulb.

Liquid conductors are used much less than the solid ones; however, they are used enough to be mentioned. For instance, the simplest water heater uses the ability of fresh water to conduct electricity by means of moving ions. This device might convert exactly all the power into heat in full accordance with the first sentence by J. Joule. Here, electrical friction is just viscous mechanical friction of liquid around ions as very small spheres. In the salty (sea) water ion concentration, electrical conductivity is much higher than in fresh water. Much less voltage is needed for conductive heating of salty water.

Gaseous conductors are mixtures of neutral particles, ion gas, and electron gas. Here, the collision of particles not only originates friction, but also splits the neutral particles, producing electrons and ions. Such a mixture is called "plasma." Ninety-nine percent of substance in the universe is in plasma state. Our sun is entirely plasma, where 4 million tons of hydrogen per second are converted into helium by thermonuclear reactions. Electrical friction is very influential on the sun's plasma behavior. Numerous plasma conductors can be seen nightly in the streets of our cities, forming announcements and ads. This plasma is called "cold ions" and neutrals do have ambient temperature and only electron gas has high temperature. Electrical friction here depends upon electron gas only.

9.1.4 CARNOT ANALOGY

Daddy, I've thought of a secret surprise
…You look just like a carp-fish with its mouth open

R. Kipling, How the Alphabet Was Made

It is in the nature of children to find and immediately indicate a similarity of different things. As usual, it does not have any serious consequences, as in the case with the daddy looking like a carp-fish. In some rare cases, the similarity (analogy) found by a clever boy had great consequences in physics.

The notorious man in the French history, Lazar Carnot, in addition to his duties as Napoleon's minister, was an excellent hydraulic engineer. He knew everything about the conversion of water energy into mechanical power by means of turbines: the power is nearly equal to water flow rate times the water height. Since his childhood, Carnot's genial son, Sadi knew this rule. He understood some similarity between the height and temperature and between the water flow mechanical power and the flow of heat. For the counterpart of the weight flow rate of water, he imagined a flow of a special liquid, which he called "caloric." In his mind, the heat flow was the caloric flow rate times temperature. The height difference of water corresponds to the temperature drop in a heat flow.

Every analogy does not mean identity. Temperature, as we know now, is measured from an absolute zero, whereas a height's zero is arbitrary. However, it does not matter when we consider differences. Unlike any other liquids, the caloric is weightless. It neither evaporates nor freezes. These minor differences did not prevent the universal acceptance of Sadi Carnot's theory. As a military engineer, he published his views in a book which in 1824 become the first textbook on heat engines. This at a time when the author was only 28 years of age. He formulated the maximal fraction of heat flow, convertible into power. It is the ratio of the temperature difference in and out of a heat engine to the absolute temperature at the entrance of heat (Carnot factor).

As the temperature at the exit of heat is not less than the ambient one, the convertible fraction is significantly less unity. To increase it, thousands of engineers after Carnot tried to increase entrance temperature in heat flow conversion. Now, the highest temperature in gas turbines is about 1700 K. If ambient temperature is approximately 300 K, the maximal fraction equals $(1700 - 300)/1700 = 0.82$.

Due to all kinds of friction, the best already attained in practical efficiency of a combined unit with gas and steam turbines is 0.60. This shows the last statement by J. Joule on the total heat convertibility to be invalid.

Let us focus now on the most unusual property of the liquid "caloric," assumed by Carnot and his contemporary colleagues: it can be born from nothing, from no substance. To produce caloric, only a friction of mechanical, electrical or thermal is needed (see below). Important transformations of the concept on caloric had happened some decades after Carnot, mainly by Rudolf Clausius. He showed that as a liquid, the caloric does not exist. He introduced the word "entropy," which became the most widely used, popular, difficult, and sometimes misleading word in physics. We do not feel much more comfortable with this word and, in the present text, prefer to keep the old word, caloric. We understand it is a step back; however, it is only one step back to make some steps forward.

Hence, as in the time of Carnot, speaking on a heat engine or a home battery, we count a heat flow as the temperature times a caloric flow rate.

9.1.5 THERMAL FRICTION

> *I am a daughter of Earth and Water*
> *And the nursling of the sky,*
> *I pass through the pores of the ocean and shores*
> *I change, but I cannot die.*

P. B. Shelley, *The Cloud*

Sadi Carnot stated in his great book that to produce maximum power from a heat flow, there should never be a temperature change without a volume change. Volume change means expansion, when temperature declines (if no heating) or compression, when it increases (if no cooling). However, in reality, there exists unavoidable heat flow if there is a temperature difference. In real life, temperature change and associated flow of heat exist everywhere. Sometimes, we try to diminish this heat flow and install a thermal insulation as we do for our home walls. Sometimes, we need to transfer thermal energy through a wall of a heat exchanger, trying to make the wall more heat conductive.

According to Carnot, every heat transfer over a significant temperature drop is the loss of our ability to do work. Why is it consistent with his analogy between temperature and height of water? Imagine that water is flowing down after a dam through a bundle of very thin pipes with a great internal surface, hence large friction. Water velocity would be low and its kinetic energy negligible, even though the total flow rate is large. Instead of pipes, imagine a sponge or another porous body, like a thick layer of sand. Here, the product of water flow rate and height equals friction work, heating the bottoming water. Potential energy of water is converted into heat. The amount of born caloric here is equal to the friction work, divided by water temperature.

A very similar process takes place in a heat transfer through a wall having a hot and a cold side. The heat flow entering the wall exactly equals the exiting one; however, the caloric flow should increase because the temperature of the exiting heat is less. Here, we see unavoidable caloric gain in a heat transfer through a wall. The more the temperature difference, the more the caloric gain. We call this phenomenon *thermal friction*. We can never distinguish the caloric born in mechanical friction from that born in the thermal one. That is why we refer to the mechanical, electrical, and thermal as generalized friction. In modern energy engineering exists very well developed methods of thermal device designs, using minimization of the sum of mechanical and thermal friction which corresponds to minimum of caloric gains.

Strictly speaking, we are obliged to indicate the fourth kind of friction known as the *chemical friction*. Every chemical reaction gives birth to some quantity of caloric. It is a more difficult part of the science on exergy and we prefer not to overburden our readers. If they would be interested, they might find an extensive discussion elsewhere.

At first glance, one may eliminate thermal friction by nearly zero temperature drops in a wall. As usual, it is misleading because the less temperature drop is the more the wall surface should be and for a large wall production, we are forced to produce much more caloric than we saved in temperature drop decline. When we deal with time-specific processes, with periodic heating and cooling, the small temperature drop leads to a very large time span due to a very slow heat transfer. There always exists an optimal temperature drop.

In every case of generalized friction, the amount of born caloric is the friction work, divided by temperature at the point. The mechanical friction work is friction force times the path, the electrical one is current times voltage drop in a conductor, and the thermal friction work in a heat transfer through a wall and is proportional to the heat flow rate times the temperature difference. If the lower temperature coincides with the ambient one, the thermal friction is maximal. Otherwise, if not in heat transfer, this fraction of heat flow, according to the Carnot analogy, might be totally converted into power. But, in heat transfer, it is completely lost for power production.

9.1.6 A WARNING

Now, we wish to define more clearly, what we are speaking about. The matter of the nature consists of the substance and the field. The latter occupies all empty space. Energy currents exist either in a substance or in a field. Energy currents in substance were discovered in 1874 by Russian physicist Nicolay Oumov. Energy currents in the field carried by electromagnetic waves were discovered by J. Pointing in 1884. These currents might be either very weak, for instance, about 10^{-15} W/sq.m from a handy phone in your pocket to about 10^{+15} W/sq.m in a powerful laser beam. The difference is in 30 orders of magnitude. The process, which might be considered as a friction in the field energy currents, has a special nature. However, it is not discussed in the present paper. We restrict ourselves within the limits of energy flows in a substance only. Every substance consists of particles, bound (solids, liquids) or unbound (gas, plasma). The interaction of particles transfers energy and creates friction.

9.1.7 RUBBER BALLOON

> *In Siberia's wastes*
> *The ice-wind's breath*
> *Woundeth like the toothed steel.*

J. C. Mangan, 1803–1849, *Siberia*

On a winter day, you carry a rubber balloon from a cold street to a warm room. Look at the balloon (Figure 9.3). It will expand and the radius of the sphere is increased until the temperatures of air inside and outside are equal. The expansion produces mechanical work of two kinds: (1) work against the rubber tension, and (2) work to shift a part of air out of the room. We assume the reader is familiar with the arithmetic of quantities, denoted by characters. The mentioned expansion process is described according to energy conservation law: the change of internal energy of the air inside the balloon (the sum of kinetic energies of air particles in their chaotic movement) is

FIGURE 9.3 Expansion work of rubber balloon in a warm room measures the exergy of air in it.

increased due to heating from room air through a rubber wall and decreased due to expansion work production. We may observe the same process when transferring a vertical cylinder with air, compressed by a heavy piston able to move in the cylinder without friction from a cold street into a warm room. When heating in the room, the pressure inside the cylinder remains constant. It is the piston weight divided by its surface. But the air expands due to temperature increase and it occupies more volume, producing some work for the piston elevation. It is the same work which in the rubber balloon is produced against rubber tension.

In both cases, according to energy conservation law, the change of internal energy of the heated portion of air ΔU, equals the external heat $\Delta Q = T_0 \Delta S$, obtained from the room with temperature T_0, less produced work $P\Delta V$. The last consists of the two parts, the work against rubber tension or piston elevation, and the work needed to shift some air from the room to outside by the room pressure P_0, so

$$\Delta U = T_0 \cdot \Delta S - (P - P_0) \cdot \Delta V - P_0 \cdot \Delta V \qquad (9.1)$$

Here U = internal energy, T = temperature, P = pressure inside the balloon or cylinder, V = its volume. Temperature and pressure in the room are indicated by subscript zero. The caloric amount is denoted by S. The reader should remember this letter for the future, when he will be forced to change the title of the symbol. The increase of caloric ΔS inside the heated balloon is caused by the heat flow from the room *and the thermal friction in the heat transfer through the rubber wall of the balloon or the wall of cylinder*. The understanding of the last part of caloric flow is of primary importance for the following definitions. It might be almost zero if the heating process is very long due to very small differences of temperature inside the balloon and in the room. In other words, if the thermal friction is absent. Actual temperature difference is as it is, but as an important assumption here, we may ignore the thermal friction and in the next formula, evaluate the maximal work done for rubber expansion or the piston elevation, specifically, the work without friction.

This work is the second term on the right-hand side of our equation. Let us denote it by A

$$A = \Delta U - T_0 \cdot \Delta S + P_0 \cdot \Delta V = \Delta(U - T_0 \cdot S + P_0 \cdot V) \qquad (9.2)$$

Our room is very large with respect to the balloon or cylinder. When they enter the room, the numbers P_0 and T_0 do not change. We may assume these as constants. That is why we transfer the difference symbol Δ outside the brackets. The temperature difference in the example is rather small, about 2°C. Even in the heart of Siberia, it does not exceed 80°C. In Siberia, outdoors it might be very cold indeed (−60°C).

9.1.8 WHAT IS EXERGY?

Man raises but time weighs.

Library of U.S. Congress ceiling sentences

Equation (9.2) for the obtained work was first formulated by the American physicist Josiah Willard Gibbs in 1873 (Gibbs, 1928). It is valid not only for a balloon, but for all cases of thermal engines. If kinetic energy or a potential energy plays a role, it might simply be added to internal one. The most important term here is the last one, reflecting the caloric gain.

The term A was first called "availability" or "available energy". Following the proposal of Z. Rant since 1956, it is referred to as *exergy*.

Exergy indeed, unlike energy, measures the ability to do work. This ability depends upon the reference state. As we have seen, energy flows always are accompanied by a friction of some kind causing energy degradation. This law of nature is no less universal and important than energy conservation law. The general definition is:

Exergy is the maximal work, attainable in given reference state without generalized friction. In the closed system, energy is conserved but exergy is destroyed due to generalized friction.

Exergy value might be attributed to a substance, like a fuel or metal or chemical compound. There are excellent tables for it, made mainly by Professor J. Szargut (1965) in Poland. Exergy flow might be described by a vector, similar to energy vector. For a given heat flow Q^*, the exergy flow is Q^* multiplied by Carnot factor $(T - T_0)/T$. When the temperature is approaching the ambient one (reference), the exergy vanishes even though the heat flow is still large.

Have a look now at Figure 9.4. It visualizes the process in a heat engine. From the left approaching the heat engine are energy units (heat). They are equal, but only a fraction of it might be elevated to the work level. A large fraction should fall to the ambient level. But without the fall, the elevation is impossible. The higher the level of coming units (temperature), the more the fraction of the work obtained. Exergy is the measure of this fraction if the heat engine is ideal (Carnot's one).

If the process temperature T is less than the ambient one T_0, the direction of exergy flow is reversed. It takes place in the heat transfer through a wall of the freezing chamber of a home refrigerator.

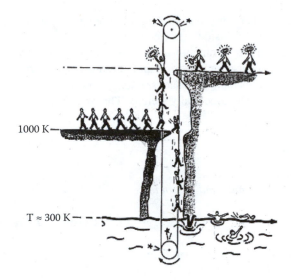

FIGURE 9.4 Conversion of heat into work in a heat engine. Maximal work is the heat exergy.

The cooling process by the use of work is presented in Figure 9.5. Work consumption lets us extract thermal energy units from the cold freezing chamber and elevate them up to ambient, thus compensating the energy units, which penetrate the thermal insulation of the chamber. Exergy of the cooled substance is the measure of needed work.

In general, in the refrigerating problems, exergy is greater than energy. It is wrong to consider exergy as a fraction of energy. Energy in a vessel with liquid helium is

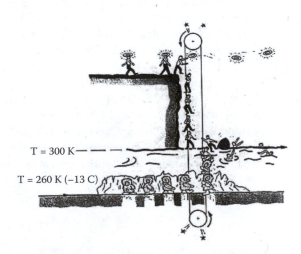

FIGURE 9.5 The work consumption for the thermal energy extraction from a freezing chamber of a refrigerator. The least work measures the exergy of coldness.

very small, whereas its exergy is quite large. It even allows power production by a turbine, if the temperature inside is low enough.

Gibbs' finding of a proper function, now called exergy, conquered all the branches of thermodynamics, energy engineering and chemical engineering. The amount of published papers reaches many thousands. Exergy now is an unavoidable term of the engineering language. The time really weighs. That is why we think it is worth your attention.

9.1.9 Reference State

A sea of stagnant idleness,
Blind, boundless, mute and motionless.

Lord Byron, *The Prisoner of Chillon*

We are such stuff. As dreams made on;
and our little life is rounded with a sleep.

W. Shakespeare, *The Tempest*

For the aforementioned balloon, the reference state (surrounding) was the air in the warm room. Its temperature and pressure do not depend upon the balloon properties anyway. In other cases when we use the word exergy, we should immediately think, what reference state is here. For many tasks, the role of the room might be played by the atmospheric air or, as in the case of a power plant, a water basin or a nearby river. In any case, the surrounding state should be something big enough to be independent from the body we are dealing with. If it is atmospheric air, we should bear in mind the seasonal and diurnal change of temperature. Exergy of ice is approximately zero in winter (by T_0 less than 0°C), but it is quite significant on a hot summer day by T_0 about 30°C; see Figure 9.5.

When we look at the aquarium with colored fish from the cockpit of a submarine, the air in the cabin is its surrounding and reference state. However, when considering the heating or cooling of this air for the comfort of the crew, the reference state is the sea water outside by pressure of many tens of bars.

In the case of a space ship, there is no surrounding substance, but a relict radiation of about 3 K temperature exists in the universe. Perhaps, it might be the reference one. For a fleet of space ships, the exergy concept has not yet been developed, and no papers have been published. This is the job of future generations of researchers.

To those readers who like difficult and rather philosophical questions, we leave you with this one: is it possible to define the exergy of the universe itself as it has no surrounding?

9.1.10 Exergy Unit

Everybody knows, a mass is measured in kg, length in m, energy in J, etc. What are the exergy units? As exergy is not energy, it is unwise to measure it in energy units of joules.

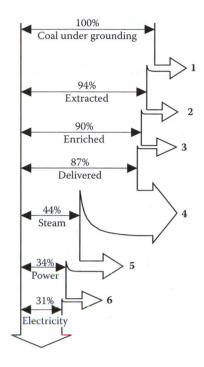

FIGURE 9.6 Exergy from a coal mine via power plant to an electricity consumer.

The offered but still not widely accepted explanation is the unit of exergy gibbs: Gi (Yantovsky, 1994). In the case of mechanical and electrical energy, 1 Gi = 1 J, but in the case of thermal or chemical processes, 1 Gi does not equal 1 J, (1000 Gi = 1 kGi, 10^6 Gi = 1 MGi, and so on). Recently, Gaggioli (2007) offered the term *Gibbs* for the entropy unit J/K. Here, we use it for exergy only to emphasize that exergy is not energy.

Look at Figure 9.6. If one assumes that exergy inflow from coal is 100 Gi, we may replace % by Gi on arrows. The fourth arrow which reflects the exergy losses in a power plant boiler should be five times less if we measure these losses in joules because energy losses in the boiler are about 9%.

Let us consider the table with figures for the comparison of steam properties before and after the steam turbine at a big power plant. The reference temperature for an exergy calculation $T_0 = 300$ K (27°C) was taken.

Temperature, K	Enthalpy, kJ/kg	Carnot Factor	Exergy, kGi/kg
900	3,600	0.67	2,400
315	2,500	0.05	125
	(69%)		(5%)

From the table is seen that completely used steam after expansion in turbines still has about 69% of its initial energy (enthalpy) before expansion, whereas its usefulness (ability to do work) is almost nil. In real economics, the price of steam needs cumbersome tables with different prices for different steam temperatures regardless of its energy content.

If the steam price is proportional to the exergy content expressed in kGi/kg, it would need only one figure X $/kGi.

9.1.11   EXERGY EFFICIENCY

Better than all measures
Of delightful sound,
Better than all treasures
That in books are found.

P. B. Shelley, *To a Skylark*

For energy management, the most important figure is efficiency: how much useful energy we have from a given energy source. If one ignores exergy and compares different units by energy efficiency only (ratio of energy output by energy input), it is often misleading and appears as efficient use, when actually it is not so. Let us compare the energy and exergy efficiencies.

System	Energy Efficiency	Exergy Efficiency
District heating boiler	0.85–1.05	0.15–0.18
Power plant boiler	0.90	0.50
Power plant	0.40	0.39
Cogeneration of power and heat	0.85	0.40
Electrical water heater	0.33	0.06
Heat pump	1.20	0.20

In all cases, the numerator is the delivered energy or exergy flow and the denominator is equal to fuel flow energy or exergy. It means that the electrical heater and electrical heat pump are considered together with a fuel-fired power plant. The numbers of energy efficiency exceeding unity are due to neglecting of vapor condensing in the boiler heating value of fuel or neglecting of low temperature heat for a heat pump case.

Some very different numbers for both efficiencies are spectacular. Both boiler cases have very high energy efficiency. Only an energy-minded engineer will never try to improve anything there. However, for an exergy-minded manager or engineer, the big thermal friction work over the temperature drop about 1000°C is evident and he really has an opportunity for improvement. When a user needs the heat of different temperatures, the best remedy against exergy destruction is cascading, heating in a series of one boiler by the flue gases and then another one with higher temperature. The exergy efficiency is the best measure indeed. It shows clearly how wisely we operate. The difference between efficiency and unity shows the exergy destruction rate.

Almost equal figures for power plant efficiencies are caused by equal numerators (electrical power is a pure exergy) and nearly equal heating value of fuel and its exergy. But nearly equal results contain the biggest difference inside. Energy efficiency erroneously shows the steam condenser as the culprit of losses, whereas the exergy correctly shows the thermal friction in the boiler. Let us try to visualize this.

9.1.12   WHERE IS EXERGY LOST?

And many more Destructions played
In this ghastly masquerade,
All disguised even to the eyes…

P. B. Shelley, *The Mask of Anarchy*

The answer is to be seen in the diagram in Figure 9.6. This visualization originated from the Irish engineer Sankey more than a century ago. He offered a graph of energy flow, with the width of a strip, proportional to the flow rate of energy in a series of energy conversion processes, one following the other. Then P. Grassmann expanded the diagram on include exergy to show where exergy is lost.

Take a look at the Sankey-Grassmann diagram for exergy flow in the conversion of coal into electricity at an ordinary coal-fired power plant in Figure 9.6. The exergy of coal in the depth is assumed as 100%. It is a national reserve, which should be consumed most efficiently.

In a coal mine with the work expenditure to crash and elevate coal, the first 6% of exergy is lost; see arrow 1. Then the coal should be enriched, i.e., the rest of the rocks should be separated. Here, some work is needed and some coal is lost, which forms the exergy losses of 4%, arrow 2. Enriched coal is delivered to the power plant, the next 3% of exergy is lost due to mechanical friction in transport, arrow 3. When the coal is burned at the power plant, we see the greatest exergy losses: 43% is lost in combustion and thermal friction, when heat is transferred from combustion gases by 1600°C to steam at 600°C (temperature drop of 1000°C), arrow 4.

Expanding steam drives the turbine coupled with the electrical generator, which produces electrical power. Here, 10% of exergy is lost due to mechanical friction in the turbine (steam over a wall, journal bearings), and electrical friction in the generator. Exergy losses in the steam condenser are also present. They are small because the temperature of condensed steam is very close to ambient (reference point). In an electric line then, about 3% of initial exergy is lost entirely due to electrical friction.

In total, only 31% of coal exergy is delivered to a consumer of electricity. Here, the power plant efficiency (34%) refers to the coal in depth. In relation to the amount of coal delivered to the site of the power plant, it is 0.34/0.87 = 0.39. Exergy losses are disguised, especially the destruction due to thermal friction. The diagram shows that the culprit of exergy destruction is the boiler of the power plants, where the losses exceed all the others, even taken together.

9.1.13   EXERGY FLOW DIRECTION

As we see, exergy is relative to energy, but as quite another function. It coincides numerically with energy when we deal with mechanical or electrical energy, but strongly differs for thermal and chemical energy. If an energy flow exists anywhere, it always is accompanied by an associated exergy. They might differ not only in size but also in direction (sign). A thermal exergy flow is always directed to a body with ambient temperature, whereas thermal energy flows toward lower temperatures. If the process temperature is higher than the ambient (reference) temperature, the direction of energy and exergy flow is the same. If we deal with a home refrigerator, cryogenics, or other cooling machine, the temperature of a flow is less than ambient and the exergy flow is opposite to an energy one. But here, history is instructive:

At the end of the nineteenth century, there were two neighbors, a brewer and a butcher. The butcher's shop was in between two brewer's rooms. In the left room, the brewer installed a cooling machine, which produced rather cold salt solution to cool a vessel during the beer preparation, located in the right room. The butcher admitted to conducting pipes with cold fluid from the left to the right rooms through his shop. When he recognized that the pipes were under a thick layer of ice, he began to store his meat around the pipes, using it as a free refrigerator. However, this immediately spoiled the beer technology and the brewer put the accident on trial before the municipal authorities as a theft. "What has been stolen?" asked the judge. "My energy," answered the brewer. But the butcher easily stated that he was not a thief of energy, but rather a sponsor of it due to the fact that energy flows from warm meat to a cold pipe, and he won the process. Had the brewer been so clever as to answer, "My exergy," he would have won the argument.

9.1.14   EXERGY FROM OCEAN

> *The sun is warm, the sky is clear*
> *The waves are dancing fast and bright...*
> *The lightning of the noontide ocean*
> *Is flashing round me...*

P. B. Shelley, *Stanzas written in dejection near Naples*

In 1881, the French physicist D'Arsonval called attention to a natural phenomenon, the difference of the temperature of the upper and lower layers of the tropical ocean's water of about 30°C above and 5°C below (deeper than 600 m). He offered to use this temperature difference for the production of power.

As we have learned already from Sadi Carnot, to produce work from heat, we need the temperature difference. In the air, the difference of 20°C is rather small and is enough to transfer heat through a wall and nothing more. But in water with high density and heat conductivity, it might be enough to produce power.

The only source of heat to upper water layers is evidently the sun. The source is abundant and clean. If we conduct a flow of upper water downward through a pipe of 1 sq.m cross-section at a velocity of 1 m/s, the flow of exergy is this: density times velocity times heat capacity times temperature difference (still it is the thermal energy flow) times the ratio of temperature difference to absolute temperature of upper water (about 300 K). The last ratio is the Carnot factor. It converts the thermal energy flow into exergy flow.

Having made the calculations, we simply get:

$$1,000 \text{ kg/m}^3 \times 1 \text{ m/s} \times 4.2 \text{ kJ/kg} \times \text{K} \times 20 \text{ K} \times 20 \text{ K/300 K}$$
$$= 5,600 \text{ kW/m}^2 = 5,600 \text{ kGi/m}^2 \times \text{s}$$

here, 4.2 kJ/kg × K is heat capacity of water and 1 kW of exergy is equal to 1 kGi/s.

We have to compare the exergy flow in the pipe with the solar energy flow density in space near Earth: 1.368 kW/m². It is called "solar constant," which is the falling energy from the sun on 1 sq.m plate, perpendicular to the light path. Falling on the ocean even in tropical regions averaged incident radiation, which was much less due to clouds and diurnal variations. It is at most 0.3 kW/m². We see that exergy flow of the ocean water is 15,000 times more dense than average radiation. It means the ocean is not only an absorber of exergy in solar light, but also the concentrator of exergy.

High exergy flow density of ocean water stimulated great activity to find a practical solution of an old D'Arsonval idea and to build a power plant, using upper ocean waters like a "fuel" to a boiler. The scheme of such floating power plant entitled OTEC (Ocean Thermal Energy Conversion) is depicted in Figure 9.7.

It is a hermetically closed loop, filled with an easy boiling substance (freon). Unlike water, this substance might evaporate and condense just at the ocean water

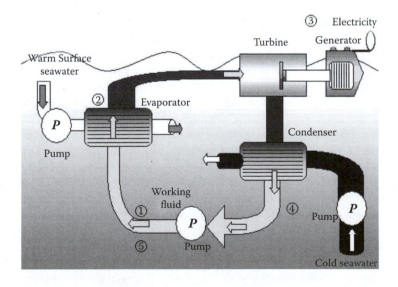

FIGURE 9.7 Scheme of OTEC aimed at consuming the highly concentrated exergy of warm ocean water for electricity generation (Xenesys Inc., http://xenesys.com/english/otec/product/index.html).

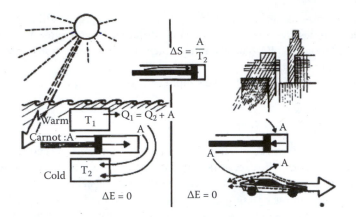

FIGURE 9.8 Exergy transfer from ocean to city by compressed air.

temperature: upper water is heating the evaporator and cold water from below cools the condenser. The liquid freon is pumped into the evaporator by elevated pressure. Inside, a loop, a turbine, and generator are installed. The unit is evidently ZEPP as it actually uses the solar energy.

There are many such experimental plants in the world, where difficult engineering work is underway to create a real engine working on the heat from the ocean. Especially active and successful are Japanese engineers.

Now, we are ready for an important "*gedankenexperiment*" as Albert Einstein referred to such a practice (see Figure 9.8).

Let us imagine that we have such an ocean engine, in which a small part of thermal energy of the upper water is converted into power and the rest of the energy in the condenser is given to the cold water below. Imagine also that near a floating power plant is floating a cylinder with air and a piston and we consumed all the generated power to shift the piston and compress the air inside. During compression, the temperature inside the floating cylinder did not increase due to heat transfer to the upper water. Exactly the same energy, which was taken in the engine as mechanical power, is returned to ocean water. There is essentially no energy change. The compressed air experiences no change either, because thermal energy of air (kinetic energy of air particles in chaotic movement) remains the same, as before compression. What really changed? The temperature profile in the ocean changed due to heat transfer, which is due to thermal friction. The exergy of the ocean has become a little less. This decline is very difficult to calculate. But gained exergy inside the cylinder is calculated very easily. It might produce work. We may take the cylinder and carry it far away from the ocean to a city.

In the city, we may imagine a car with a pneumatic engine, working on compressed air. We may install our cylinder and move for a significant distance. Produced work against mechanical friction in the movement is converted into heat by ambient temperature, which exactly covers the cooling of air due to its expansion in the pneumatic engine. There are no changes in energy, but exergy of the air in the cylinder is completely lost by the wheels driving the vehicle. Here, we see the production of exergy, its transportation and use: a car movement. It is very difficult, if not impossible, to explain what really happened in the ocean and the city using only energy terms. Energy did not change either in the ocean or in the city.

9.1.15 HEATING OF DWELLINGS

She picked out a nice dry Cave… and she lit a nice fire of wood at the back of the Cave, and she hung a dried wild horse skin, tail-down, across the opening of the Cave.

R. Kipling, *The Cat that Walked by Himself*

In the Stone Age, when people used to make their dwellings (cave) warm by a fire inside, almost all the heat produced remained in the dwelling. However, in addition to heat, there remained unpleasant combustion products. This was changed a thousand years after by decoupling fire, gases, flow, and dwelling space, thus eliminating unpleasant breathing of combustion products. But this drastically declined the heating system efficiency. Stack gases took away a lion's share of produced heat. Since old ancient times, the amount of fuel spent to heat dwellings has greatly increased.

In such northern countries as Canada, Russia, Scandinavia, and in colder years in Britain, the fuel consumption for heating dwellings is the major part of the national energy balance. It is important to understand that almost all the dwelling heating is thermal friction and exergy destruction.

We wish to avoid any cumbersome calculations. Imagine that thermal insulation of boiler houses and water pipes is perfect and heat flow Q^* is constant from a boiler to a home battery.

The flue gases average temperature is about 1100 K and the reference temperature (air outside) is 270 K. The fraction of Q^*, the exergy flow equals $(1100 - 270)/1,100 = 0.75\ Q^*$. In a home battery, the temperature is admitted at most 90°C or 363 K. Here, the exergy flow is $(363 - 270)/363 = 0.25\ Q^*$. If to remember the exergy destruction in combustion and some hydraulic friction the mentioned figures of exergy efficiency $0.15 - 0.18$ are evident, it means *only* one sixth of fuel exergy can reach the consumer. A much smaller fraction of the fuel exergy, about 6%, can reach the home room when we use an electrical heater.

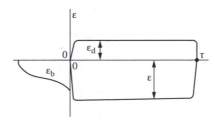

FIGURE 9.9 The history of an energy unit, including construction and operation time (Yantovsky, 1984).

The first man who understood the great possibility of fuel saving in heating was William Thomson (much later, Lord Kelvin). In 1852, he published the paper "On the Economy of Heating or Cooling of Buildings by Means of Currents of Air." There the idea of heat multiplier was described. It is comprised of an air expander, a heat exchanger, a compressor and a heat engine. When ambient air expands to about 70% of atmospheric pressure, it cools down, and then in the heat exchanger, it is heated by an ambient air and then compressed to the normal pressure. Its temperature is increased over ambient. It is enough to be comfortable in a dwelling even when it is very cold outside. Note that the fuel is used not to heat air directly, but to feed the engine, driving the compressor. Instead of great thermal friction, the temperature increase is created by an expansion–compression process in accordance with Carnot's advice to avoid heat transfer.

William Thomson optimistically promised to heat houses with 3% of the fuel used in contemporary stoves. We may wonder how close this figure is this to the exergy efficiency of the former (inefficient) heating systems. This idea gave birth to a big business in the world, where some tens of millions of such machines, referred to as heat pumps, exist. They differ from Thomson's scheme by using many other substances instead of air and being driven mostly by an electric motor. In many cases of big units, drive by heat engines, either piston or turbine, are also used. No heat pump can create a big temperature difference. The more it is used, the less is the fuel economy. Figure 9.10 (heat pump) is lost. The still useless energy units from ambient water (low-grade heat) are elevated to a higher temperature by means of much less noble work units, coming from the left. The useful energy, going to the right, is the sum of low-grade heat and consumed work. We have seen already in the table of efficiencies a comparison that exergy efficiency of a heat pump exceeds a district heating one and is three times more than that of an electric water heater.

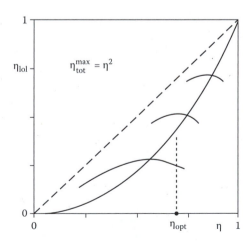

FIGURE 9.10 Total efficiency versus exergy efficiency (Iantovski, 1998).

The challenge to future heat supply engineers is to achieve a figure of attainable exergy efficiency of about that of a power plant, which would mean 40% instead of today's figure of not more than 20%. It will give tremendous fuel savings and be economically justified.

9.1.16 THE MAGIC NUMBER

The history of the World is the biography of great men.

Library of U.S. Congress

First, we wish to show our observations, when looking at the greatest achievements of classical thermodynamics as the basis of its exergy branch. We found the year of publishing of the most eminent contributions and the year of the authors' births. We find the following:

	S. Carnot	R. Mayer	R. Clausius	W. Thomson	N. Oumov
Contribution published in	1824	1842	1850	1852	1874
Author birth year	1796	1814	1822	1824	1846
Age at publication	28	28	28	28	28

Looking at this magic number 28, we used to tell our readers: hurry up. You don't have too much time. For the contemporary matriculates, it is less than a decade.

Now, we try to answer the questions of the curious boy: energy is conserved. The flow of energy in an insulated channel is constant. Exergy disappears due to generalized, mostly thermal, friction. It is just exergy we need to produce and save. The rate of its destruction depends on us and on our skill to manage it. Exergy drives all the wheels on Earth. Just exergy supports life. The exergy century has come. Fight thermal friction!

If any readers are interested in learning more on the subject, there are a number of books and publications listed below which may be enlightening:

Bejan, A. 1997. *Advanced Engineering Thermodynamics*. New York: J. Wiley & Sons, Inc.

Borel, L. 1988. *Thermodynamique, Énergétique et Mutation dans l'Évolution du Vivant*. Ecole Polytechnique Federal de Lausanne, Dept. De Mecanique. Lausanne: EPFL (in French).

Hammond, G. and A. Stapleton. 2001. Exergy Analysis of the United Kingdom Energy System. *Proc. Inst. Mech. Engrs.* Part A (215): 141–162.

Häfele, W. 1976. "Science, Technology and Society—A Prospective Look," presented at Bicentennial Symp. on Science: A Resource for Humankind? Nat. Ac. Sci., Oct. 11–13, Washington, D.C., U.S.A.

Szargut, J., Morris, D. and F. Steward. 1988. *Exergy Analysis of Thermal, Chemical and Metallurgical Processes*. New York: Hemisphere.

Szargut, J. 2005. *Exergy Method. Technical and Ecological Applications*. Southampton-Boston: WIT Press.

Yantovsky, E. 1994. *Energy and Exergy Currents*. New York: NOVA Sci. Publ.

There also is a professional journal: *EXERGY — An International Journal*, ISSN: 1164-0235, published by Elsevier, which might be of interest to the reader.

9.2 EXERGONOMICS

Nomenclature

a	correlation parameter
C	correlation parameter
D	sum of exergy destructions
d	wall thickness
F	surface area
J	electrical current
j	current density, investment cost
K	net exergy coefficient
k	overall heat transfer coefficient
L	length
m	correlation parameter
Q	heat flow
q	heat current density
T	absolute temperature
t	temperature drop
Z	main exergonomic criterion
α	T_1/T_0, temperature ratio
β	parameter
δ	exergy current density
ε	exergy (or its flow with a dot)
τ	normative time
ξ	specific invested exergy
ρ	mass density
η	efficiency

Subscripts

1,2	sides of a wall
0	reference state
opt	optimal
d	delivered
con	consumer

9.2.1 Exergy versus Money

Exergonomics is a mirror image of ordinary economics, using only exergy expenditures instead of monetary ones. Some examples of optimization by a simple relation of invested exergy and current exergy expenditures—including heat transfer through a wall, an electrical conductor, and a thermal insulating wall—are given.

The synthesis of thermodynamics and economics have a great history. The early part of the history has been described by Gaggioli and El-Sayed (1987) and Tsatsaronis (1987), who proposed the title "exergoeconomics" at the beginning of the 1980s. More recent papers are published in the *Proceedings of International Conference* with an acronym ECOS, and are respected throughout the world. The detailed description of exergoeconomic methodology involving the cost of exergy expenditures in monetary units has been given by Tsatsaronis (1987) and in the textbook (Bejan, Tsatsoronis, and Moran, 1996). This methodology is widely used by many practitioners in real designs.

However, the same practitioners are uneasy about the price uncertainty. As a well-known example, oil prices might be mentioned. Within the time span of the design and construction of an ordinary major power plant, in 1973 to 1983, oil prices quadrupled and then dropped to their original value. Imagine the quandary of a designer in 1973. Which prices then should he use in his design? At the same time, exergy expenditures were much more firm and stable. For instance, the data (Szargut and Morris, 1987) for construction materials and fuels are stable and valid for a long time. That is why, aside from the well-known ExergoECOnomics. the much less-known branch of exergy analysis, referred to as "exergonomics" (Yantovsky, 1989, 1994) was developed. It is a mirror image of ordinary economics but with the value of every good or service measured in exergy expenditures instead of money. These expenditures are divided on two major parts: invested and current.

The approach goes back to Frederick Soddy, who tried to replace money by means of energy expenditures at the beginning of this century (Soddy, 1926). The crucial role of exergy, rather than of energy, is now evident.

The use of restrictions of Second Law in cogeneration calculations and cost was clearly advocated by G. Arons (1926):

> … the evaluation of rejected heat not by amount of calories, but by power which it may give passing through an ideal steam engine …

Then he recommended multiplying the heat flows by the Carnot factor, which is exactly what we now do in the exergy accounting of cogeneration power plants.

The formulation of a target function for exergy accounting in power plant optimization (including the invested and current parts) was presented in DeVries and Nieuwlaar (1981, p. 367]:

> … before any lower bound … has been reached, one bounces the minimum of the total exergy requirement … exergy of capital and fuel/heat.

DeVries and Nieuwlaar (1981) mentioned Soddy's attempt and its systematic elaboration by Chenery (1949). The presentation of total specific exergy expenditure as the sum of inversed exergy efficiency and net-exergy coefficient with its minimization in general form might be found in Iantovski (1997, 1998). It is also reproduced in this paper.

While exergy expenditure alone may by no means be the basis for decision-making, neither may money. Real life forces us to take into account pollution, time restrictions, and other factors. That's why exergonomics might be considered as an auxiliary tool to prove the solutions made on other bases. For more reliable decision making, the simultaneous optimization by at least three target functions (exergy-money-pollution) is needed. It is a more difficult task, not discussed here.

At present, it is an accepted fact that no decision in engineering is possible without a computer simulation of the problem. The more the computer is involved, however, the more the logic of the solution remains vague. The computer never takes the place of logic. This is especially true in education, where some logic should be transferred to the students. For educational purposes, simplified problems should be constructed which admit a simple straightforward solution.

The paper is aimed at presenting a clear model of a link between the invested and current exergy expenditures, which lets us find the optimal exergy efficiency, and some simple examples of exergonomic optimization, recommended for educational purposes.

9.2.2 THE MAIN CRITERION OF EXERGONOMICS

When considering exergy efficiency, we should clearly define what our system is, where are its boundaries of the entrance and the exit, and then divide the outlet (delivered) exergy flow by the inlet (consumed) flow. If our system actually is a subsystem or a part of a greater system, our partial optimization might not coincide with the optimum of the greater system. By comparison of fuel-fired plants, the input exergy flow should be fuel. It is not obligatory if only a part of the unit is optimized.

The total life story of every energy conversion unit is presented in Figure 9.9, where ε_B = invested exergy needed to manufacture the unit. The total exergy efficiency is the ratio of delivered exergy to the sum of exergy expenditures:

$$\eta_{tot} = \frac{1}{(1/\eta + 1/K)} \tag{9.1}$$

where

η is an ordinary exergy efficiency, based on current exergy flows
$K = \varepsilon \, (d\tau)/\varepsilon_B$ is net exergy coefficient, ratio of delivered exergy to invested exergy

The inverse quantity

$$Z = \frac{1}{\eta_{tot}} = \frac{1}{\eta} + \frac{1}{K} \tag{9.2}$$

is the main criterion in exergonomics in the same sense as COE (cost of energy) is the main criterion in economics, which is usually subjected to minimization.

This approach is similar to the so-called ELCA (exergy life cycle assessment); see for example Szargut (1971). However, the word "exergonomics" appeared in Corneliessen (1997) a little earlier and the subject differs from ELCA by exergy discounting requirements and in some other respects, not mentioned in present book.

9.2.3 INVESTED EXERGY MODELS

In general, for the arbitrary function $K(\eta)$, the optimal solution as seen by Yantovsky (1989) and Corneliessen (1997) is:

$$Z_{\min} = \frac{1}{K}\left[1+\left(-\frac{dK}{d\eta}\right)^{1/2}\right]; \qquad \eta_{opt} = K \bigg/ \left(-\frac{dK}{d\eta}\right)^{1/2} \qquad (9.3)$$

which is found by solving $\partial Z/\partial \eta = 0$. The problem consists of finding the particular function $K(\eta)$, reflecting every case study.

The first attempt to find correlation between investment cost and exergy efficiency belongs to J. Szargut (1971). He selected for the monetary investment cost j, the function

$$j = j_0 \frac{\eta}{1-\eta} \qquad (9.4)$$

which gives the right trend, $j \to \infty$ as $\eta \to 1$.

The current approach to cost evaluation of capital investment through exergy expenditures is described by Bejan et al. (1996), which also lists many additional references.

An approximation with the two correlation parameters a and m has been proposed by Iantovski (1998):

$$K = a \cdot \eta^{-m} \qquad (9.5)$$

which gives

$$Z_{\min} = \left(\frac{m}{a}\right)^{\frac{1}{m+1}} + \frac{1}{a}\left(\frac{a}{m}\right)^{\frac{m}{m+1}}; \qquad \eta_{opt} = \left(\frac{a}{m}\right)^{\frac{1}{m+1}} \qquad (9.6)$$

The deficiency of the model, Equation (9.5), is its inability to reflect the main physical condition of the approach of exergy efficiency to one, the infinite increase of the size of equipment and, hence, its invested exergy.

Here, the other model is recommended as rather simple and relevant: the invested exergy is proportional to the delivered exergy and inversely proportional to the exergy destruction rate

$$\varepsilon_B = \frac{\varepsilon_d \cdot \tau}{C^2(1-\eta)} \qquad (9.7)$$

Here, C^2 is a single correlation parameter needed to adjust Equation (9.7) to any particular problem. The needed trend $\varepsilon_B \to \infty$ as $\eta \to 1$ is evident. From Equation (9.7), we obtain:

$$K = \frac{\varepsilon_d \tau}{\varepsilon_B} = C^2(1-\eta) \tag{9.8}$$

$$Z = \frac{1}{\eta} + \frac{1}{K} = \frac{1}{\eta} + \frac{1}{C^2(1-\eta)} \tag{9.9}$$

$$\frac{\partial Z}{\partial \eta} = \frac{1}{\eta^2} + \frac{1}{C^2(1-\eta)^2} = 0; \quad \eta_{opt} = \frac{C}{1+C}; \quad Z_{min} = \left(\frac{1+C}{C}\right)^2 \tag{9.10}$$

$$\eta_{tot}^{max} = \eta_{opt}^2 \tag{9.11}$$

If Equation (9.7) is valid, the maximum total efficiency is equal to the square of the optimal one. Equation (9.11) seems to be the shortest formula in exergy analysis. In Figure 9.10, one can see the curves η_{tot} versus η and the geometrical place of maxima, forming the square parabola. The three superimposed curves reflect the increase of C.

9.2.4 DC ELECTRICAL CONDUCTOR

This case is of interest because it is the simplest example with explicit invested exergy, a complete correspondence of exergy to power. Entropy does not play any role here. Let us consider a direct current power line with given current J, length L, cross-section area F, made of material with the electrical conductivity σ, density ρ, and exergy intensity ξ (specific exergy consumption to produce 1 kg of material). The sum of the fuel exergy losses due to electrical resistance and invested exergy to build the conductor is

$$D = \frac{1}{\eta} J^2 \frac{L\tau}{\sigma F} + \xi \rho L F \tag{9.12}$$

From $\partial D/\partial F = 0$, we obtain

$$j_{opt} = \frac{J}{F_{opt}} = \left(\frac{\xi \rho \sigma \eta}{\tau}\right)^{1/2} \tag{9.13}$$

For an aluminium conductor, $\xi = 330$ MJ/kg, $\rho = 2500$ kg/m^3, $\sigma = 2.5 \times 10^7$ Ohm $\times$ m^{-1}, $\tau = 9 \times 10^{10}$ s (33 years), $\eta = 0.35$, and the optimal current density is 0.085 A/mm^2. It is much less than in common practice (about 1 A/mm). Note here the power plant efficiency η, which reduces electrical exergy to that of fuel.

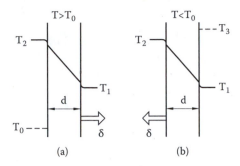

FIGURE 9.11 Heat transfer through a wall for (a) a power plant ($T > T_0$) and (b) refrigerator ($T < T_0$).

9.2.5 HEAT TRANSFER THROUGH A WALL

Two fluid streams of temperatures T_2 and T_1 are separated by a wall of thickness d; see Figure 9.11.

The total heat flow through the surface F is:

$$Q = F \cdot q = F \cdot k \cdot (T_2 - T_1) = F \cdot k \cdot t; \qquad t = (T_2 - T_1) \qquad (9.14)$$

The exergy current density from the left is

$$\delta_2 = q(T_2 - T_0)/T_2 = k \cdot t \cdot (T_2 - T_0)/T_2 \qquad (9.15)$$

The exergy current to the right is

$$\delta_1 = q(T_1 - T_0)/T_1 = k \cdot t \cdot (T_1 - T_0)/T_1 \qquad (9.16)$$

T_1 is less than T_2, therefore, δ_1 is less than δ_2. Their difference is merely the exergy destruction. The exergy efficiency of the heat transfer process is

$$\eta = \frac{\delta_1}{\delta_2} = \frac{(T_1 - T_0)}{(T_2 - T_0)} \cdot \frac{T_2}{T_1} \qquad (9.17)$$

When $t \to 0$, we have $\eta \to 1$; however $q \to 0$ and $F \to \infty$ for any given Q.

Assuming that the wall is flat, or that its curvature radius is much greater than the wall thickness, the invested exergy is

$$\varepsilon_B = Fd\rho\,\xi \qquad (9.18)$$

The delivered exergy for a normative time is

$$\varepsilon_d = F\delta_1\,\tau = Fkt\tau(T_1 - T_0)/T_1 \qquad (9.19)$$

while the consumed exergy is

$$\varepsilon_{con} = F\delta_2\tau = Fkt\tau(T_2 - T_0)/T_2 \tag{9.20}$$

The net exergy coefficient is

$$K = \frac{\varepsilon_d}{\varepsilon_B} = \frac{kt\tau}{d\rho\xi}\frac{(T_1 - T_0)}{T_1} = \beta t\frac{(T_1 - T_0)}{T_1}; \qquad \beta = \frac{k\tau}{d\rho\xi} \tag{9.21}$$

Finally, the main exergonomic criterion is

$$Z = \frac{(T_1 - T_0)}{T_1}\left(1 - \frac{T_0}{T_1 + t} + \frac{1}{\beta t}\right) \tag{9.22}$$

From $\partial Z/\partial t = 0$, we have $(T_1 + t)^2 = \beta T_0 t^2$, and optimal temperature drop

$$\frac{t_{opt}}{T_1} = \left[\pm(\beta T_0)^{1/2} - 1\right]^{-1} \tag{9.23}$$

Denoting $\alpha = T_1/T_0$, for maximal total efficiency, we have

$$\eta_{tot}^{max} = (\beta T_0)^{1/2}\cdot\left\{[(\beta T_0)^{1/2} - 1]\cdot\left[1 + \frac{1}{(\beta T_0)^{1/2}(\alpha - 1)}\right] + \frac{\alpha}{(\alpha - 1)}\right\}^{-1} \tag{9.24}$$

The behavior of maximal efficiency η and t_{opt} is presented in Figure 9.12. It is clear that a high total efficiency is possible only for a sufficiently high value of the governing dimensionless criterion.

For a more concrete analysis, let us introduce the fictitious temperature drop

$$t_f = \frac{1}{\beta} = \frac{d\cdot\rho\cdot\xi}{k\cdot\tau} \tag{9.25}$$

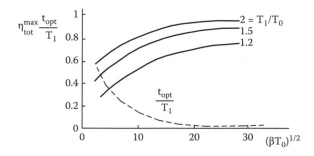

FIGURE 9.12 Maximum total efficiency versus the governing criterion $(\beta T_0)^{1/2}$, and optimal temperature drop versus the same criterion (Iantovski, 1998).

This is needed to transfer the heat flow $d\rho\xi/\tau$, to supply the energy embodied in the wall for the time τ. Now the criterion

$$(\beta T_0)^{1/2} = (T_0/t_f)^{1/2} \tag{9.26}$$

takes the form of the root of the ratio of reference temperature to the fictitious drop.

The numerical data for a shell-and-tube counter-flow high pressure air heater with inlet temperature 450 K, $d = 1$ cm, $k = 50$ W/m² K, $\xi = 100$ MJ/kg for steel tube, $\rho = 1 \times 10^4$ kg/m³, and life time of 10 years by 5500 hours/year are

$$(\beta T)^{1/2} = 17.3, \quad \alpha = 1.5, \quad t_{opt} = 27.6 \text{ K}, \quad \eta = 0.816$$

If high pressure air is heated up to 600 K, then $\alpha = 2$, $t = 36.8$ K, and $\eta = 0.90$.

So far, we have considered the case $T > T_0$. Another case is refrigeration ($T < T_0$); cf. Figure 9.11. Heat flows as before, from left to right; however, the exergy current is reversed. Repeating the preceding analysis, we have

$$\eta = \frac{(T_0 - T_2)}{T_2} \cdot \frac{T_1}{(T_0 - T_1)}; \quad K = \frac{(T_0 - T_2)}{T_2} \cdot \frac{k\tau}{d\rho\xi} \cdot t \tag{9.27}$$

The main criterion is

$$Z = \frac{T_2}{(T_0 - T_2)}\left(\frac{T_0}{T_2 - t} - 1 + \frac{1}{\beta t}\right) \tag{9.28}$$

From $\partial Z/\partial t = 0$, we have

$$t_{opt} = T_2 \cdot \left[1 \pm (\beta T_0)^{1/2}\right]^{-1} \tag{9.29}$$

The minus sign in Equation (9.29) makes no physical sense, because $t > T_2$ is impossible. For a typical refrigerating recuperator at 180 K, made of aluminum tubes with $\rho = 3000$ kg/m³, $d = 2$ mm, $\xi = 300$ MJ/kg, $k = 1000$ W/m² K and $\tau = 10$ years, we have $(\beta \times T)^{1/2} = 180$, and $t_{opt} = 1$ K, which is in agreement with common practice. Note, that k is the overall heat transfer coefficient, which includes the thermal resistance of the wall itself (d/λ) and convective heat transfer on the both surfaces, α_1 and α_2.

9.2.6 THERMAL INSULATION OPTIMIZATION

An absolute thermal insulation is impossible. Heat leakage exists as soon as a temperature difference exists. Sometimes, the engineering task is to diminish this leakage regardless of expenditure, as is the case in attempts to reach absolute zero. However, in industry and architecture, it is important to determine the reasonable level of thermal insulation. For us, the word "reasonable" means the least sum of exergy lost through the thermal insulation, and exergy spent to manufacture this insulation; see Figure 9.12.

Here the specific heat current is

$$q = k \cdot (T_2 - T_0) \tag{9.30}$$

The exergy current lost through insulation is

$$\delta = q \cdot \frac{(T_2 - T_0)}{T_2} = k \cdot \frac{(T_2 - T_0)^2}{T_2} \tag{9.31}$$

The overall heat transfer coefficient is

$$k = \left(\frac{1}{\alpha_1} + \frac{d}{\lambda} + \frac{1}{\alpha_2} \right) \tag{9.32}$$

The sum of the lost exergy and the invested exergy is

$$D = \delta\tau + d\rho\xi \tag{9.33}$$

$$\frac{\partial D}{\partial d} = -\frac{(T_2 - T_0)^2}{\lambda T_2} \cdot \frac{\tau}{(1/\alpha_1 + d/\lambda + 1/\alpha_2)^2} + \rho\xi = 0 \tag{9.34}$$

$$d_{opt} = \frac{\lambda(T_2 - T_0)}{(\lambda T_2 \rho\xi/\tau)^{1/2}} - \lambda\left(\frac{1}{\alpha_1} + \frac{1}{\alpha_2} \right) \tag{9.35}$$

$$q_{opt} = \left(\frac{1}{\tau} \lambda T_2 \rho\xi \right)^{1/2} \tag{9.36}$$

The most important application of these results seems to be in civil engineering: how to select the optimal wall thickness. Here are the two numerical examples:

Brick wall for a building: $T_2 = 295$ K, $T_0 = 273$ K, $\lambda = 1$ W/m × K, $\rho = 2{,}640$ kg/m³, $\xi = 5$ MJ/kg, $\tau = 50$ years, $q = 49.7$ W/m², $0.24 < d_{opt} < 0.44$ for $10 < \alpha_{1,2} < 1000$ W/m² × K.

Thermal insulation of glass wool for a dry ice (solid carbon dioxide) storage, $T_2 = 200$ K, $T_0 = 295$ K, $\lambda = 0.03$ W/m × K, $\rho = 50$ kg/m³, $\xi = 30$ MJ/kg, $\tau = 40$ years.

Here the results $q_{opt} = 2.67$ W/m² and $d = 1.06$ m, do not depart from the common practice.

We see now that it is possible to develop a routine procedure to find the trade-off between invested exergy and current exergy expenditure. This procedure can be described analytically in very simple terms, and can be used for a very sophisticated 3-D optimization of reduced emissions units.

9.3 EXERGY CONVERSION IN THE THERMOCHEMICAL RECUPERATOR OF A PISTON ENGINE

9.3.1 EXAMPLE OF EXERGY CALCULATION

Here is presented an example of exergy calculation in a rather complicated case of a zero emissions piston engine, described in Section 7.4.

Thermochemical recuperation (TCR) as the use of waste heat for the fuel reforming with H_2O or CO_2, leading to efficiency increase, is well known for gas turbine units. Less known is that the similar TCR is useful for piston engines (PE) as well. The difference is in the low pressure of the mixture of fuel and H_2O or CO_2.

This section is aimed at calculating endothermal reactions of methane as fuel with H_2O and CO_2 by conditions, suitable for PE. It is shown that exergy of heat of exhaust gases is converted into exergy of reformed fuel (CO + hydrogen).The results are presented as graphs of exergy efficiency versus temperature of the end of reaction. Within the limits of 800 to 1000 K, the exergy efficiency of TCR is increased from 0.23 to 0.77 for CH_4 and H_2O and from 0.30 to 0.80 for CH_4 + H_2O. Fuel economy in a diesel engine might reach 19%.

The initiative to apply to piston engines the principle of thermochemical recuperation, well known from gas turbine systems, belongs to the U.S. company, Gas Technology Institute. The literal citation (Gas Technology Institute, 2004) is used:

> TCR is a promising concept for application to reciprocating engines. Rejected sensible energy from the engine combustion is recovered to support catalytic reformation of the fuel (natural gas) to a higher energy content fuel.
>
> Preliminary analysis by GTI indicates that a TCR system could offer the following advantages over conventional power generating systems:
>
> Reduced fuel consumption by virtue of increased cycle efficiency, which results from use of rejected heat from the engine
> Reduced CO_2 emissions by virtue of increased efficiency
> Increased power output capacity
> Ultra-low emissions of criteria pollutants (NO_x, CO, etc.) and hazardous pollutants (HCs, formaldehyde, etc.) by virtue of improved combustion conditions
>
> Theoretically, the TCR can reform all kinds of liquid and gaseous hydrocarbon fuels.
> The hydrogen-enriched fuel would increase the ignitability of the mixture due to its broad flammability limits and low-ignition energy requirements, and increase flame speed relative to natural gas.

From an energy balance point, the TCR is a clear measure to decline the temperature of outgoing gases (e.g., losses) by virtue of absorption of their energy by incoming gases.

In addition to GTI statements, we wish to underscore that TCR is promising for the efficiency increase in zero emissions (including automotive) piston engine systems (Yantovsky et al., 2004), where fuel should be reformed not with steam, but with carbon dioxide.

In application to gas-turbine systems with CO_2 capture and TCR, the detailed analysis, demonstrating the exergy losses in gas turbine, fuel reactor, etc., has been made by Bolland and Ertesvag (2001).

This paper is aimed at calculating the exergy balances of TCR for a piston engine to visualize the losses of the reformer and select the needed reaction temperature by reforming with water and carbon dioxide as well. For simplicity, the only gaseous fuel to be considered is methane.

9.3.2 PROCESSES IN TCR

The two endothermic reactions of fuel methane reforming are known:

$$CH_4 + H_2O \implies CO + 3H_2, \tag{9.37}$$

and

$$CH_4 + CO_2 \implies 2CO + 2H_2 \tag{9.38}$$

Schematics of a TCR are presented in Figure 9.13, where 1, 2, 3 and 4 are the flows of incoming reactants, 5 is the flow of reaction products, 6, 7, 8, 9, and 10 are flows of exhaust gases of piston engine, giving its sensible heat to support endothermic reactions.

The equilibrium constant K by atmospheric pressure depends upon the temperature only and relates to the change of standard Gibbs energy ΔG_T^0:

$$K = \exp\left(-\frac{\Delta G_T^0}{RT_i}\right) \tag{9.39}$$

For the first reaction, Equation (9.37):

$$\Delta G_T^0 = 188,748 + 228.75 \cdot T_i - 69.61 \cdot T_i \cdot \ln T_i + 80.711 \cdot 10^{-3} \cdot T_i^2/2$$
$$- 22.874 \cdot 10^{-6} \cdot T_i^3/6 \tag{9.40}$$

and for the second reaction, Equation (9.38):

$$\Delta G_T^0 = 230,323 + 193.78 \cdot T_i - 71.75 \cdot T_i \cdot \ln T_i + 106,81 \cdot 10^{-3} \cdot T_i^2/2$$
$$- 36,06 \cdot 10^{-6} \cdot T_i^3/6 \tag{9.41}$$

The fractions of reformed methane until equilibrium for the two reactions are

$$m = \sqrt{\frac{\sqrt{k}}{2,6} \Big/ \left(1 + \frac{\sqrt{k}}{2,6}\right)}, \quad \text{and} \quad m = \sqrt{\frac{\sqrt{k}}{2} \Big/ \left(1 + \frac{\sqrt{k}}{2}\right)}. \tag{9.42}$$

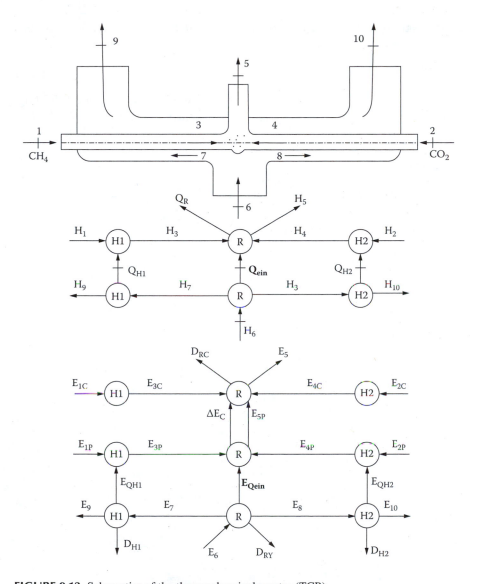

FIGURE 9.13 Schematics of the thermochemical reactor (TCR).

The fractions of reaction products for two reactions, respectively, are:

$$\mu_{CH_4} = \mu_{H_2O} = \frac{1-m}{2\cdot(1+m)}; \quad \mu_{CO} = \frac{m}{2\cdot(1+m)}; \quad \mu_{H_2} = \frac{3\cdot m}{2\cdot(1+m)} \qquad (9.43)$$

$$\mu_{CH_4} = \mu_{CO_2} = \frac{1-m}{2\cdot(1+m)}; \quad \mu_{CO} = \mu_{H_2} = \frac{m}{(1+m)} \qquad (9.44)$$

The ratio of the amount of final reaction products to the incoming reactants (chemical coefficient of molecular change) is

$$\beta = 1 + m \tag{9.45}$$

The incoming reactants are at ambient temperature $T_1 = T_2 = T_0$. Due to heating, their temperature increases. It is assumed that the reaction is getting started by $T_3 = T_4 = T_{AR}$, when $m = 0.001$. In the first reaction, $T_{AR} = 415$ K, whereas in the second one $T_{AR} = 515$ K.

This heating energy is:

$$Q_{H1} = H_2 - H_1 = H_7 - H_9, \qquad Q_{H2} = H_4 - H_2 = H_8 - H_{10}$$

Reactions in the point R take the energy:

$$Q_{in} = H_6 - H_7 - H_8 = Q_R + (H_5 - H_3 - H_4)$$

9.3.3 EXERGY BALANCE

Exergy flow from cooled flue gases is:

$$\Delta E_{CQ} = E_{Qein} + E_{3P} + E_{4P} - E_{5P} \tag{9.46}$$

or per one mole

$$\Delta e_{cq} = e_{qein} + e_{3p} + e_{4p} - e_{rp} \tag{9.47}$$

$$e_{qein} = q_{ein} \cdot \eta_c = q_{ein} \cdot \left(1 - \frac{T_0}{T_i}\right)$$

$$q_{ein} = q_R + (h_{r5} - h_3 - h_4)$$

$$a_M = \sum \mu_i \cdot a_i; \quad b_M = \sum \mu_i \cdot b_i; \quad C_M = \sum \mu_i \cdot C_i$$

$$h_{r5} = 2 \cdot \beta \cdot h_5 = 2 \cdot \beta \cdot \left(a_M + \frac{b_M}{2} \cdot T_i + \frac{C_M}{3} \cdot T_i^2\right) \cdot T_i$$

$$e_{rp} = 2 \cdot \beta \cdot e_{5p} = 2 \cdot \beta \cdot [(h_5 - h_0) - T_0 \cdot (s_5 - s_0)]$$

Molar exergy of the flow of a substance consists of two parts (see Figure 9.13):

$$e_i = e_{ip} + e_{ic} \tag{9.48}$$

where $e_{ip} = (h_i - h_o) - T_o(s_5 - s_o)$ is the physical exergy, depending upon the pressure and temperature of flow and e_{ic} = chemical exergy, depending on contents. In Table 9.1, the standard chemical exergy, recommended by Szargut and Petela (1965) is given.

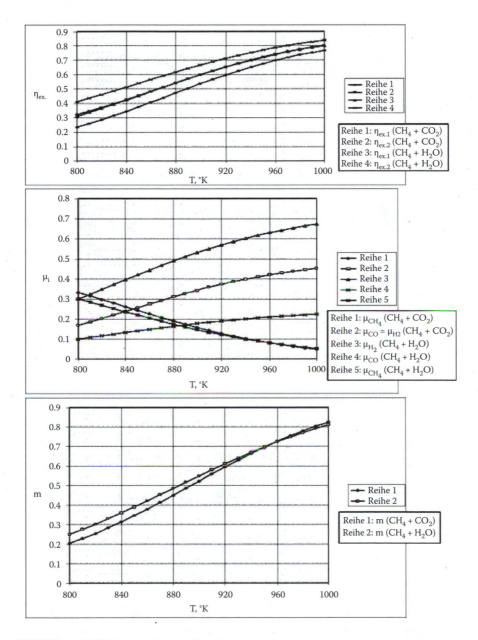

FIGURE 9.14 Efficiency and molar fraction versus temperature.

In an endothermic reaction, the thermal exergy is converted into exergy of substances and reaction losses:

$$\Delta e_{cq} = \Delta e_{c} + d_{RC} \tag{9.49}$$

TABLE 9.1
Reaction Heat and Molar Exergy of the Fuel Constituents (at 298 K, 1 Bar)

Substance	h_{high} kJ/kmole	h_{low} kJ/kmole	e_i kJ/kmole
H_2	285,900	241,800	238,350
CO	283,000	283,000	275,350
CH_4	890,500	802,300	836,510
CO_2	—	—	20,170
H_2O	—	—	750
O_2	—	—	3,970
N_2	—	—	720

where

$$\Delta e_c = e_{rc} - e_{3c} - e_{4c} \tag{9.50}$$

$$e_{rc} = 2 \cdot \beta \cdot e_{5c} = 2 \cdot \beta \cdot \Sigma \mu_i e_i$$

and

$$e_{q_{ein}} = \Delta e + d_{RC}$$

$$\Delta e = e_5 - e_3 - e_4, \quad e_5 = e_{rp} + e_{rc} \tag{9.51}$$

$$e_3 = e_{3p} + e_{3c}; \quad e_4 = e_{4p} + e_{4c}$$

Exergy efficiency of the TCR is defined as follows:

$$\eta_{ex.1} = \Delta e_c / \Delta e_{cq} = 1 - d_{RC} / \Delta e_{cq} \tag{9.52}$$

TABLE 9.2
Contents of Reformed Fuel by 1000 K

Reaction	CH_4	H_2O	CO	H_2
$CH_4 + H_2O$	0.052	0.052	0.224	0.672
$CH_4 + CO_2$	0.0479	0.0479	0.4521	0.4521

TABLE 9.3
Exergy Conversion Versus Temperature (e in 10^5 kJ/kmole)

Ti, K	$\Delta e_{cq} = \Delta e_c + d_{RC}$		Δe_c		η_{ex1}	
	(1)	(2)	(1)	(2)	(1)	(2)
515	0.835	0	0.00402	0	0.005	0
550	0.85439	1.04996	0.00933	0.00431	0.0109	0.0041
600	0.94599	1.14313	0.02518	0.01386	0.0266	0.0120
650	1.01952	1.23710	0.05813	0.03736	0.0570	0.0302
700	1.0925	1.3177	0.11984	0.08756	0.1097	0.0664
750	1.1647	1.3962	0.2239	0.18293	0.1922	0.1310
800	1.2402	1.4780	0.3822	0.3452	0.3082	0.2335
850	1.3207	1.5671	0.5954	0.5879	0.4508	0.3751
900	1.4043	1.6627	0.8392	0.8909	0.5976	0.5358
950	1.4843	1.7556	1.0677	1.1868	0.7193	0.6759
1,000	1.5546	1.8365	1.243	1.4088	0.7995	0.7671

To relate the gained chemical exergy in reformed fuel to the physical one from flue gases, we get

$$\eta_{ex2} = \Delta e / (\Delta e + d_{RC}) \tag{9.53}$$

9.3.4 RESULTS OF CALCULATIONS

The results of computer calculations of endothermic reactions and exergy conversion efficiency versus the temperature of a reaction are given in Table 9.3 and Figure 9.14. It is observed that in the reforming with CO_2 in reaction (9.38), the exergy taken from exhaust gases is greater than that by reforming with H_2O. But exergy converted into the new fuel exergy in the second reaction is less than in the first one. That is why the efficiency η_{ex1} for reaction (9.37) is higher. This difference vanishes when the temperature exceeds 900 K.

If needed, the end of the reaction might be by temperature less 1000 K. There are no rapid changes in the curves contents versus temperature.

In the graphs in Figure 9.14, the conversion of methane in reformed fuel in the temperature interval of 800 to 1000 K is presented. By $T = 1000$ K, the exergy efficiency may reach 80%.

In order to evaluate the possible increase of exergy of reformed fuel, let us look at the figures calculated for a gas-fuelled diesel engine of about 5000 kW capacity. The main exergy flows are in Table 9.4.

The chemical exergy flow which enters the cylinder is about 19% higher than that of methane fuel because of the use of exergy of flue gases. As exergy efficiency of the diesel itself only depends slightly upon used fuel, it shows the possible fuel economy by TCR of about 19%.

TABLE 9.4
Exergy Flows

Exergy	Fuel, E_3	Reformed, E_5	Flue Gas, E_6	Loss, D_R	Outlet Gases, E_7	E_8
Total, kW	96,157.7	11,880.48	3,245.93	66.42	571.48	604.22
Temperature	(515 K)	(1,000 K)	(1,102 K)	—	(586 K)	(586 K)
Chemicals	9,588.0	11,393.58	629.04	—	305.76	323.28
%	100	118.83	27.29	—	—	—

The percentage distribution of exergy, taken from flue gases $E_{Qein} = 2070.2$ kW is as follows: 20.75% is used for heating reformed fuel from 515 up to 1000 K, 76.05% is converted into chemical exergy of reformed fuel, and 3.2% is exergy destruction (internal losses due to entropy rise in chemical reaction).

Calculations show a possibility of reaching rather high exergy efficiency (about 80%) of TCR incorporated in the piston engine systems when reaction temperature is 900 to 1000 K. Methane reforming by CO_2 is only slightly less efficient than reforming by water. Fuel economy by TCR in a gas-fuelled diesel is about 19%.

9.4 CURRENTOLOGY AS AN INTERMEDIATE FILE

9.4.1 DIVERGENCE FORM EQUATION

Nomenclature

A	magnetic potential
B	magnetic induction
D	electrical induction
E	electrical field
e	specific energy
F	cross-section
I	information (amount)
I*	flow of information
i	information in unit of volume
J	current density vector
G	chemical potential
H	magnetic field
h	specific enthalpy
k	Boltzmann constant
L	length
M	magnetization
P	polarization
P	power
p	pressure

q	specific electrical charge
S	entropy (amount)
S^*	entropy flow
s	specific entropy
T	temperature
U	internal energy
$\mathbf{V}$	velocity
v	$1/\rho$ = specific volume
W	work
$\mathbf{e}$	exergy
α	thermoelectric coefficient
γ	electrical conductivity
δ	energy current vector
δ_{ik}	tensor unity
ε	electrical permeability
μ	magnetic permeability
ρ	mass density
σ	entropy gain intensity
τ_{ik}	tangential stress tensor
ϕ	electrical potential
ω	rotational frequency

Subscripts

i	information
q	electrical charge
s	entropy
ik	tensor components
o	vacuum or reference

Note: Vectors are printed in boldface type, vector product is denoted by an ×, scalar product of vectors is denoted by a •.

Every energy process is the interaction of different flows. Even in a steady-state system, where nothing depends upon time, there exist steady-state currents. It reflects the unit watt, which is the current of one joule per second.

In this section, we attempt to describe flows of mass, impulse, electrical charge, energy, entropy exergy, and information in the standard divergence form of differential equations for the vectors of their currents. Entropy and information are considered as positive and negative thermal charges. Some numeric examples are given. Currentology might be a useful intermediate file in the array of different branches of energy engineering, especially with respect to the zero emissions power units.

In energy engineering education, thermodynamics, hydrodynamics and electrodynamics are the firm basements. As usual, they are thought of differently by different lecturers and as a result, students do not necessarily see what they have in common.

There exists a possibility to implement a new introduction, which now, in the twenty-first century, looks like an intermediate file in a computer storage of basic information of the engineering science.

It is based on the presentation of conservation laws in the divergence form of a differential equation for a quantity Q:

$$\partial Q / \partial t + div\,\mathbf{J} = g \tag{9.54}$$

Here, $\mathbf{J}$ = vector of the current of Q, g = gain or sinc (negative gain) of the Q. If g is equal to zero, Equation (9.54) means a conservation law.

The Gauss theorem for steady-state flow is well known as the total quantity of Q, born in the given volume V equals the flow of vector $\mathbf{J}$ through the surface F, surrounded V. It means

$$\iiint_V g\,dV = \iint_F (\mathbf{J} \bullet n)\,dF \tag{9.55}$$

Now to select a particular case, the particulate vector $\mathbf{J}$ and the specific gain g should be found. The Q might be mass, impulse, electrical charge, energy, entropy, and exergy.

Equations (9.54) and (9.55) describe transport processes in hydrodynamics, thermodynamics, and electrodynamics.

In the middle of the nineteenth century, classical mechanics had been made, whereas the heat theory and electrodynamics were at their infancy. At that time, the main efforts of many physicists were directed to mechanical models of electrical and thermal phenomena. Today, electrodynamics is ahead of all due to the universal validity of Maxwell's equations and very wide applications area. It is worthwhile to try to develop some electrical models of thermal processes for energy engineering. They include the flows of mass, impulse, electrical charges, energy, and entropy. In the last decade, the flows of exergy and information have been added.

As it travels on flows or currents, no assumption of equilibrium in ordinary thermodynamic sense is valid. These topics are described by nonequilibrium thermodynamics (NeT) when the local equilibrium takes place and even in a gas flow, it is possible to identify the definite pressure and temperature in a point of a flow. Known books by de Groot and Mazur (1983), Landau and Lifshits (1982) and Sutton and Sherman (1965) are good examples of different flow descriptions. Such books are useful for every reader regardless of the professional skill.

Over the last decades, this author has tried to find useful analogies for thermal and electrical engineering (Yantovsky 1989, 1997a, 1997b). In the section, a further development of the same line is presented as an introduction of a branch of the NeT dealing with currents and flows entitled as *Currentology*.

Along with entropy and exergy currents, it is useful to include the information flow as a current of a charge which is treated on the base of the concept of thermal charge of both signs. Admitting the thermal charge generation, one should get rid of any mechanical models.

9.4.2 INFORMATION AS NEGATIVE ENTROPY

Leo Szilard (1929) (as cited in Leff and Rex, 1990) stated that the creation of one bit of information always is accompanied by $k\ln 2$ of entropy. It was the starting point of thermodynamics of information similar to origination of classical thermodynamics by Sadi Carnot. Recently, many authors further developed this discovery.

Modern Wikipedia cites the statement of G.N. Lewis about chemical entropy in 1930: "Gain in entropy always means loss of information and nothing more." And then in Wikipedia: "…possession of a single bit of Shannon information (a single bit of negentropy in Brillouin's term) really does correspond to a reduction of physical entropy, which theoretically can indeed be converted into useful physical work" (http://en.wikipedia. org/wiki/Entropy).

Brillouin (1956, in Leff and Rex, 1990):

Every physical measurement requires a corresponding entropy increase, and there is a lower limit, below which measurement becomes impossible. This limit corresponds to change in entropy of… kln2 or approximately 0.7k for one bit of information obtained.

Here I should stress that there exist other views, neglecting Brillouin's approach, especially for biological problems (Khazen, 1991).
Zemanski (1968) stated:

A convenient measure of information, conveyed when the number of choices is reduced from W_o to W_1 is given by $I = k \times \ln(W_o/W_1)$. The bigger the reduction, the bigger the information. Since $k \times \ln W$ is the entropy S, then $S_1 = (S_o - I)$, which can be interpreted to mean that the entropy of a system is reduced by the amount of information about the state of the system… The increase of information as a result of compression is seen to be identical with the corresponding entropy reduction.

Weinberg (1982, in Leff and Rex, 1990) opined:

We can quantify the information per measurement or per bit, it is $I = -k \ln 2$, the negative sign meaning that for each increase in information there is a decrease in actual physical entropy.

Machta (1999) stated:

Students are now comfortable with the notion that information is physical and quantitatively measurable… definite amount of information may be stored in digital form on hard drives and other storage media and in dynamic memory… the information content of a record is the number of bits (ones or zeros) needed to encode the record in the most efficient possible way. This definition is formalized by algorithmic information theory… Suppose that the hard drive is initially filled with a record which is the result of 8 billion coin tosses. The entropy, associated with information-bearing degrees of freedom will be $k \times (8 \times 10^9)$ bits. Suppose that disc is erased… To satisfy the Second Law, an equal or greater increase of entropy must have occurred… a tiny amount of heat $k \times (8 \times 10^9) \times (300 \text{ K}) = 2.3 \times 10^{-11}$ J must be released.

According to the known Landauer principle "there is a minimum dissipation of kT whenever a bit of information is erased in an environment at temperature T."

Schlögl (1989) observed: "... the development of the microstate in time is not exactly known. The description is restricted to probabilities over microstates. The information associated with their distribution cannot increase after the last observation. In thermal states, this information is a function $I(M)$ of the state variables. Consequently, there exists the function which has the mentioned features of entropy:

$$S(M) = -k \cdot I(M) \qquad (9.56)$$

This identification of macroscopic entropy with the lack of information about the microstate is the basic link between statistics and phenomenological thermodynamics."

Schlögl's statement is published in the book by the same publisher as R. Clausius *Abhandlungen über die mechanische Wärmetheorie,* Vieweg, Braunschweig.

Frankly speaking, we should look at the works of rare authors rejecting any link of information and thermodynamic entropy, but the majority admit that relations between information entropy and thermodynamic entropy have become common currency in physics.

9.4.3 THERMAL CHARGES

All the students easily accept the electrical energy transfer by conduction of electrical charges through a metallic conductor. Here, electrical charge is an energy carrier. The conductive transfer of thermal energy we call heat has another carrier, the entropy.

We define this process as conduction, because the substance in which entropy flows, might be at rest. Similar to electrical current, the entropy current takes place relative to a substance. Regardless of the possible statistical interpretation of entropy, it plays the role of a charge flow in the conduction process.

When electrical space charge is transferred by a moving insulator, it is a convective electrical current. A similar process is the convective transfer of thermal energy by flow of hot water. More comments on the conduction and convection division will follow in a separate section.

As we have seen that entropy and information have the same units and differ by sign only, we may treat them as the thermal charge of different signs. They are additive, extensive quantities, their potential (intensive quantity) is temperature. Along with the elemental electrical charge 1.6×10^{-19} Coulomb, there exists the elemental thermal charge 1 bit $= k \times \ln 2$ (about 1×10^{-23} J/K). Every bit of information by measurement needs not less than $k \times \ln 2$ entropy production. A measurement might be treated as a kind of splitting of a neutral particle, forming a pair of charges but thermal instead of electrical. When meeting the thermal as well as electrical, charges annihilate if they are of opposite signs.

A sharp contrast exists between the mentioned charges: between electrical ones exists a strong force, creating the field (Coulomb force and Lorentz force when it moves in magnetic field), whereas between thermal charges, there are no repulsive or attractive forces and no field.

The most important difference is the strict conservation law for electrical charge (when by splitting of a neutral particle, the positive charge is exactly equal to the negative one) and the lack of this analogy for thermal charges. Due to the Second Law in every real process, the positive charge is to be in excess. Only in ideal (reversible) processes, thermal charges of both signs are equal, which means the thermal charge conservation for ideal processes.

9.4.4 Generalized Friction

The first observation of electrical charges in ancient times was due to mechanical friction of some pieces of different materials (amber = electron in Greece). Here, the splitting of neutral surface molecules takes place by a kind of rubbing. The positive thermal charge always is created by friction, not only mechanical. Yantovsky (1994) proposed the concept of generalized friction, including mechanical, electrical, and thermal friction. Mechanical friction does not need any explanation, electrical is the result of scattering and collisions of electrons in a conductor (electrical resistance and Joule heating) and, most important, thermal friction is the heat flow over the significant temperature drop. Mixture of different gases or liquids might be associated with chemical friction. To reverse the mixing, significant work is needed. Generalized friction produces positive thermal charge. There exists, however, a possibility to observe some intensive positive thermal charge flow with creating negative ones (producing order from a chaos as in Prigogine's dissipative structures).

In general, friction creates only a positive thermal charge, whereas to produce a negative charge, work is needed for a separation process. Generalized friction is the cause of irreversibility. The appeal of F. Boshniacovich: "Fight irreversibilities," in Russian translates to "fight friction."

The notorious negentropy of Schrödinger and Brillouin, which is needed to feed living creatures, is just the negative thermal charge, which is neutralized in excess by metabolic processes. Here, like in every combustion reaction, the chemical part of generalized friction takes place.

9.4.5 Some Equations

Most problems in energy engineering are described by equations, valid for continua, neglecting relativistic and quantum effects, and assuming local thermodynamic equilibrium. The last means the equilibrium in a volume large enough with respect to mean free path of a gas particle and small enough in comparison with a channel size. It is just our assumptions for currentology, as in ordinary gas dynamics with definite T and P.

The mass conservation equation is:

$$\partial \rho / \partial t + div\, \mathbf{V} = 0 \tag{9.57}$$

Electrical charge conservation has a similar form

$$\partial q / \partial t + div\, \mathbf{J}_q = 0 \tag{9.58}$$

The non-conservation of the thermal charges of both signs is described by the two equations

$$\partial s/\partial t + div\, \mathbf{J}_s = \sigma_s \qquad (9.59)$$

$$\partial i/\partial t + div\, \mathbf{J}_i = \sigma_i \qquad (9.60)$$

with the short formulation of the Second Law $\sigma_s \geq \sigma_i$.

The sum of Equations (9.59) and (9.60), when the latter is multiplied by k gives

$$\partial(s - k \cdot i)/\partial t + div(\mathbf{J}_s - k \cdot \mathbf{J}_i) = \sigma_s - k \cdot \sigma_i \qquad (9.61)$$

The energy conservation law in the standard divergence form is as follows:

$$\partial e/\partial t + div\, \delta = 0 \qquad (9.62)$$

The components of energy *in a substance* are described in an extended Gibbs identity:

$$de = dU + T \cdot d(s - ki) - p \cdot dv + \varphi \cdot dq + \mathbf{H} \bullet d\mathbf{M} + \mathbf{E} \bullet d\mathbf{P} \qquad (9.63)$$

Here, the shear part of mechanical work is omitted, and compression work increases when volume decreases and i itself is considered positive.

A substance, where all the terms of Equation (9.63) are of comparable magnitude, is unknown. If a compressible gas is under consideration, there is no polarization, magnetization or information flow. A solid body, however, might be magnetized or polarized and carry a large amount of information (CD, DVD, sticks, and their forthcoming heirs). Here, some processes of energy conversion might be depicted on *T-I* diagrams, similar to ordinary *T-S* diagrams (in which $S > 0$). Instead of isobar or isochor on *T-I* diagrams, the isofield or isopolarization lines are to be used with the possibility of thermodynamic cycles drawing.

9.4.6 IMPULSE CONSERVATION

As a quantity which obeys the strict conservation law (Newton's law), impulse is much more complicated than mass or electrical charge. We distinguish the mechanical impulse (per unit of volume) $\rho\mathbf{V}$, which coincides with the mass current vector, and electromagnetic impulse

$$\mathbf{D} \times \mathbf{B}, \quad \text{where} \quad \mathbf{B} = \mu_0 \mathbf{H} + \mathbf{M}, \quad \mathbf{D} = \varepsilon_0 \mathbf{E} + \mathbf{P} \qquad (9.64)$$

A complicated case is the interaction of substance and field, when a body has an impulse, being at rest ($\mathbf{V} = 0$), when it is polarized and magnetized. It might be true for some new materials as magnetodielectrics or magnetic liquids (colloids of magnetics in a dielectric liquid).

The two parts of the impulse current are mechanical P_{ik} and electromagnetic M_{ik} currents of impulse. Both are tensors of second rank:

$$P_{ik} = -p\delta_{ik} + \tau_{ik} + \rho V_i V_k \tag{9.65}$$

$$M_{ik} = -E_i D_k + H_i B_k - \delta_{ik}(\mathbf{D} \bullet \mathbf{E} + \mathbf{B} \bullet \mathbf{H}) \tag{9.66}$$

For each part of impulse, the standard divergency form equations of currentology are as follows:

$$\partial(\rho \mathbf{V})/\partial t + divP_{ik} = q\mathbf{E} + \mathbf{J}_q \times \mathbf{B}, \tag{9.67}$$

$$\partial(\mathbf{D} \times \mathbf{B})/\partial t + divM_{ik} = -(q\mathbf{E} + \mathbf{J}_q \times \mathbf{B}) \tag{9.68}$$

The rhs of these equations is the intensity of impulse creation or destroying, which is just the force density (force in unit of volume). The force is the link between the mechanical and electromagnetic parts of impulse. For the total impulse, the conservation law is as follows:

$$\partial(\rho \mathbf{V} + \mathbf{D} \times \mathbf{B})/\partial t + div(P_{ik} + M_{ik}) = 0 \tag{9.69}$$

Well-known effects of mechanical impulse conservation such as rocket flight, aircraft, or marine propulsion are familiar from school days. Less known are the effects of the total impulse conservation such as solar light pressure (first measured by P. Lebedev) giving some hopes for space navigation by solar sails. Much higher pressure creates the beam of powerful laser. The forging of plastic metals by a high current discharge or an induction electric motor torque are other examples.

9.4.7 ENERGY CONSERVATION

N.A. Oumov in 1874 was the first, presented energy conservation equation as the divergence one:

$$\frac{\partial e}{\partial t} + \left[\frac{\partial(e\mathrm{V}_x)}{\partial x} + \frac{\partial(e\mathrm{V}_y)}{\partial y} + \frac{\partial(e\mathrm{V}_z)}{\partial z} \right] = 0 \tag{9.70}$$

At that time, there were no symbol div. Nevertheless, it is evident, that (9.70) is just a divergence-form equation of currentology. In 1884, J. Pointing, using Maxwell's field equations, gave the energy equation for electromagnetic field:

$$\frac{\partial}{\partial t}(\varepsilon_0 \mathbf{E}^2 + \mu_0 \mathbf{H}^2)/2 + div(\mathbf{E} \times \mathbf{H}) = -\mathbf{E} \bullet \mathbf{J}_q \tag{9.71}$$

As usual in a substance, the Ohm's law is valid: $\mathbf{J}_q = \gamma \times \mathbf{E}$ and the rhs takes the form $-\mathbf{J}^2/\gamma$. In this case, the field energy in a closed system should always decrease.

The sign of rhs might be reversed if there exists an electromotive force, greater than the field **E**, when the Ohm's law is

$$\mathbf{J}_q = \gamma(\mathbf{E} - \mathbf{E}_{mf}) \tag{9.72}$$

Mechanical energy conservation for a compressible, viscous, thermally, and electrically conducting substance is as follows:

$$\frac{\partial}{\partial t}\left[\left(U + \frac{p}{\rho} + G + \frac{V^2}{2}\right)\rho + \frac{1}{2}(\varepsilon_0\mathbf{E}^2 + \mu_0\mathbf{H}^2)\right]$$

$$+ div\left[\left(U + \frac{p}{\rho} + G + \frac{V^2}{2}\right)\cdot\rho V + V \bullet \mathbf{P}_{ik} - \lambda \cdot grad\, T\right] = \mathbf{E} \bullet \mathbf{J}_q \tag{9.73}$$

The sum of Equations (9.71) and (9.73) gives the total energy conservation equation, where rhs is equal to zero:

$$\frac{\partial}{\partial t}\left[\left(U + \frac{p}{\rho} + G + \frac{V^2}{2}\right)\rho + \frac{1}{2}\left(\varepsilon_0\mathbf{E}^2 + \mu_0\mathbf{H}^2\right)\right]$$

$$+ div\left[\left(U + \frac{p}{\rho} + G + \frac{V^2}{2}\right)\cdot\rho V + V \bullet \mathbf{P}_{ik} - \lambda \cdot grad\, T + \mathbf{E} \times \mathbf{H}\right] = 0 \tag{9.74}$$

Here, the substance is assumed as nonmagnetized and nonpolarized. The Oumov-Pointing vector, describing the energy current, is in the square brackets. From Equations (9.71) and (9.74), we learn that it is defined by its divergence only. It means that the addition to it of any solenoidal vector **a** (div**a** = 0) does not affect the equations. Such a vector after Slepian (1942) has been $rot(\phi\,\mathbf{H})$ and the Pointing vector **E** × **H** was transformed into a more convenient form

$$\delta = \mathbf{E} \times \mathbf{H} + rot(\varphi\mathbf{H}) = \varphi(\mathbf{J}_q + \partial\mathbf{D}/\partial t) - \partial\mathbf{A}/\partial t \times \mathbf{H} \tag{9.75}$$

By the definition

$$rot\mathbf{A} = \mathbf{B}, \quad \mathbf{E} = -grad\varphi - \partial\mathbf{A}/\partial t.$$

For a steady-state case, when $\partial(\ )/\partial t = 0$, the modified Pointing vector is

$$\delta = \varphi\cdot\mathbf{J}_q \tag{9.76}$$

and we see that electrical current vector lines coincide with energy vector ones. For a gas flow, which often is a carrier of energy, introducing the entropy current vector

$$T\cdot\mathbf{J}_s = \rho\mathbf{V}(U + p/\rho) - \lambda \cdot grad\, T \tag{9.77}$$

we have the steady-state case of energy current in a substance

$$\delta = (G + V^2/2) \cdot \rho \mathbf{V} + \mathbf{V} \bullet \tau_{ik} + \varphi \cdot \mathbf{J}_q + T \cdot \mathbf{J}_s \qquad (9.78)$$

In vacuum, the only energy carriers are electromagnetic waves. There, no one term of Equation (9.78) is acting.

9.4.8 EXERGY CURRENT VECTOR

Exergy, the ability to do work in the given reference state, is the main concept of modern energy engineering. There exists a large amount of textbooks on the matter. The first one was written by Szargut and Petela (1965) and more recently by Szargut in 2005.

The specific exergy is

$$e = G_\in + V^2/2 + U + p/\rho + \mathbf{H} \bullet \mathbf{B}/\rho + \mathbf{E} \bullet \mathbf{D}/\rho - T_o(S - k \cdot I) \qquad (9.79)$$

G. Wall (1977) cited the pioneering works by M. Tribus, C. Bennett, and R. Landauer and stated after M. Tribus: "... relation between exergy e in Joule and information I in binary units (bits) is $\mathbf{e} = k' \times T_0 \times I$, where $k' = k \times \ln2 \approx 1 \times 10^{-23}$ J/K." This early statement is valid if exergy is totally associated with negentropy (the term $-T_0 \times S$). As is evident from Equation (9.79), there exist many other important terms.

In Equation (9.79) as before, we count $S > 0$, and $I > 0$. The sum of positive and negative thermal charge times reference temperature T_0 represent the lost or gained (Iantovski, 1997a) works. Instead of chemical potential, the chemical exergy $G_\in$ should be used. The divergence form of exergy equation in currentology is

$$\partial e/\partial t + div \mathbf{J}_\in = -T_0 \cdot (\sigma_s - k\sigma_i) \qquad (9.80)$$

in agreement with Equation (9.61). Introducing entropy current in Equation (9.77) into the corresponding exergy current vector, we have

$$\mathbf{J}_\in = \rho \mathbf{V} \cdot (G_\in + V^2/2 + \mathbf{H} \bullet \mathbf{M} + \mathbf{E} \bullet \mathbf{P})$$

$$= \mathbf{E} \times \mathbf{H} + rot(\varphi \mathbf{H}) = \varphi(\mathbf{J}_q + \partial \mathbf{D}/\partial t) - \partial \mathbf{A}/\partial t \times \mathbf{H} + \mathbf{V} \bullet \tau_{ik} + (T - T_0)(\mathbf{J}_s - k \cdot \mathbf{J}_i)$$

$$(9.81)$$

The simplified steady-state version for a gas flow is (Iantovski, 1997b):

$$\mathbf{J}_\in = \rho \mathbf{V} \cdot (G_\in + V^2/2) + \varphi \mathbf{J}_q + \mathbf{V} \bullet \tau_{ik} + (T - T_0) \cdot \mathbf{J}_s \qquad (9.82)$$

The last term in the case of thermal conduction or convection translates into $(\lambda \times \operatorname{grad} T) \times (1 - T_0/T)$ or $(\rho \, VU) \times (1 - T_0/T)$ not only in a gas, but in liquids, and the solid bodies.

9.4.9 Conductive, Convective and Wave Transfer

In every textbook on heat transfer. one may find the above mentioned division for thermal energy transfer. Kreuzer (1984) offered to use conduction and convection division for an impulse flow. Looking at Equations (9.81) and (9.82), we see the transfer by moving substance ρV convection and flows through a steady-state substance (conduction). This division is valid for all the terms of energy and exergy currents, and is very useful for currentology. In energy equations, all the quantities, carried by convection, are the function of the state, not of the path. That is why it is wrong to call the convection of internal energy U as heat. It is just the thermal energy transfer. The functions of a path, the work, and the heat are all the *energy transfer conduction* either by impulse (it is work) or by entropy (it is heat); see Yantovsky (1989). The first (Oumov) energy conservation equation in divergency form Equation (9.70) is just energy convection by moving substance. The hot water flow in a heating system is a convection of thermal energy from boiler to user, whereas the conductive transfer of this thermal energy through a wall of home battery is heat.

The third modes of energy and exergy transfer are waves, which can carry impulse, entropy and information. The most important waves are the electromagnetic ones, predicted by Maxwell and discovered by Herz. These waves (radio, TV, magnetron oven, light, laser) carry the energy current $\partial \mathbf{A}/\partial t \times \mathbf{H}$. For the unit of this current might be used the solar constant $\varnothing = 1{,}368$ W/m^2. It is the perfectly measured light energy current from the sun falling on an Earth satellite in space.

In the next figures, the numerical examples of exergy current densities are measured in $\varnothing$:

Natural gas main, $\rho V G_{\in}$	$\approx 3.3 \times 10^6\ \varnothing$
Hot water heating, $(\rho\, VU) \times (1 - T_0/T)$	$\approx 3{,}000\ \varnothing$
Boiler furnace, $(\rho\, VU) \times (1 - T_0/T)$	$\approx 7{,}000\ \varnothing$
Boiler tube wall, λ grad $T \times (1 - T_0/T)$	$\approx 150\ \varnothing$
Electrical generator gap, $\partial \mathbf{A}/\partial t \times \mathbf{H} \approx \omega BLH$	$\approx 180{,}000\ \varnothing$
High voltage direct current line, $\phi\, \mathbf{J}q$	$\approx 7 \times 10^8\ \varnothing$
Wind by 10 m/s, $\rho V \times V^2/2$	$\approx 0.36\ \varnothing$
Averaged solar in Europe	$\approx 0.11\ \varnothing$

These figures vividly show why an energy boiler should be much bigger in size than an electrical generator. Looking at very low figures for popular renewables (wind and solar), one may understand why the bulk energy supply by solar or wind energy is hardly possible in densely populated, energy intensive, but rather small land surface countries of Europe. It is because in an ordinary fuel-fired power plant, the exergy current density is thousands of times more than that of solar or wind. It means they need much, much less land. The question seems to be crucial for the most densely populated and energy intensive part of Europe, the land of Nordrhein Westfalen in Germany. V. Smil (1991) systematically investigated the energy current density in energetics and indicated the possible problems of renewables due to very small energy current densities and much land needed.

Figures of another kind illustrate the thermal charges currents. The known, rather old, figures of the attained in practice information flows, measured in bits/s are as follows:

Telegraph: 75; telephone: 2500–8000; television: 2×10^7; glass fiber: 1×10^8.

Let us compare the currents of thermal charges in an optical glass fiber line. Imagine a cable, connecting a warm room (27°C) with a cold one (0°C), possessing a length of 27 m. Here, the grad(T) = 1 K/m. The entropy conduction flow is:

$$S^* = \mathbf{J}s \times F = F \times (\lambda \times \text{grad}T)/T = 4.5 \times 10^{-11} \text{ W/K, by}$$
$$\lambda = 1.34 \text{ W/m.K}, F = 0.01 \text{ mm}^2, T = 300 \text{ K}.$$

The negative charge flow (information) is:

$$k \cdot I^* = 1.38 \cdot 10^{-23} \cdot 10^8 = 1.38 \cdot 10^{-15} \text{ W/K}$$

it is four orders of magnitude less than the positive one. If, however, the recent data on the achieved information flows are used, the figures are different. Modern optical lines may carry 320 Gbit/s (Bishop et al., 2001); here $k \times I^* = 4.4 \times 10^{-12}$ W/K which is very near to mentioned S^*. The new technology of MEMS (Micro-electro Mechanical Systems) demonstrated in July 2000 more than 10 Terabits per second of total switching capacity. Such switches "might support the petabit (quadrillion-bit) systems that are not very far over the horizon." It is evident that soon the flows of information might exceed the thermal conductivity entropy flows in optical fibers.

Let us compare the information flow from a computer and entropy income due to conversion of electrical power into heat inside it. This ratio might be considered as the thermodynamic computer efficiency: $\eta = k \cdot I^*/S^* = T_0 \cdot k \cdot I^*/P = 8.4 \cdot 10^{-14}$.

Here it is assumed I^* as for TV, $P = 100$ W, $T_0 = 300$ K. For a telefax machine by $I^* = 9,600$ bit/s and $P = 20$ W, we have $\eta = 2 \times 10^{-18}$. For a modern notebook computer by $P = 1$ W and information output of 10 Gb/s, this efficiency is 4×10^{-11}.

The quite natural question here is this: are the figures with such a small efficiency meaningful? Could they make any sense? The answer is: probably yes. The visible way to increase this efficiency was mentioned by Richard Feynman in his lectures on the theory of computation: to shift from a crystal to molecular level in the chip structure and to shift to low power consuming transmitters of information. If a handheld telephone transmits 8000 bit/s, its power supply should be not less than: $T_0 \times k \times I^* = 300 \times 1.38 \times 10^{-23} \times 8000 = 3.3 \times 10^{-17}$ W.

Contemporary devices consume much more power and the limit is still not within sight. The trend, however, is toward it.

9.4.10 INFOELECTRIC EFFECT EXPECTATION

Let us consider Ohm's law, Equation (9.72), with the thermoelectric electromotive force:

$$\mathbf{J}_q = \gamma \cdot (\mathbf{E} - \alpha \cdot grad\, T) \qquad (9.83)$$

From here, the field is

$$\mathbf{E} = \mathbf{J}_q/\gamma + (\alpha \cdot T/\lambda)(\mathbf{J}_s - k \cdot \mathbf{J}_i) \qquad (9.84)$$

As the main assumption here along with entropy current $(1/T) \times \lambda \times \mathrm{grad}\ T$, the information current was included. The reason is a possible similarity of interaction of thermal charges of both signs with the conductive electrical charge flow.

This expected effect has been entitled "infoelectric," analogous to well-known thermoelectricity (Iantovski and Seifriz, 1996). In isothermal condition, when a wire is immersed in a bath, beside an electrical current, there exists only the information current:

$$\mathbf{E} = \mathbf{J}_q/\gamma + k \cdot T \cdot \mathbf{J}_i \cdot \alpha/\lambda \qquad (9.85)$$

The ratio of the second term of rhs to the first one gives a criterion D_e:

$$D_e = k \cdot T(\alpha \cdot \gamma/\lambda) \cdot I^* /(\mathbf{J}_q \cdot F) \qquad (9.86)$$

The more D_e, the more pronounced is the expected effect. To observe it in addition to the careful choice of wire material (figure of merit $\alpha\gamma/\lambda$), the ratio of information flow to electrical current is important. In other words, how many bits of thermal charge carries one coulomb? If $T = 300$ K, $\alpha = 0.001$ V/K, $\gamma = 1000$ kSm/m, $I^* = 100$ Mbit/s, $\lambda = 60$ W/mK, and $\mathbf{J}_q \times F = 1$ mkA, the criterion $D_e = 0.0006$. The expected potential difference should exceed the noise due to thermal motion of electrons. In an experiment, the two identical wire probes with the same electrical current might be used, assembled as shoulders in a sensitive bridge scheme. When the information current appears in one of them, the bridge should be disbalanced.

9.5 PARETO OPTIMIZATION OF POWER CYCLES

Nomenclature

c,m,p	given correlation parameters from statistical data
K	net-exergy coefficient
M	monetary target function
P	pollution target function
r	specific pollution (per exergy unit)
u,v,w	priority coefficients for exergy, money, pollution
Z	exergy target function
ε	exergy or exergy flow
λ	specific cost (per unit of exergy)
τ	lifetime of considered object operation
η	exergy efficiency (η^* = Pareto-optimal, η_{tot} = total)

Subscripts

d	delivered
con	consumed
b	invested

9.5.1 COORDINATES FRAME

Here, the three-dimensional space is formed by the three quantities: money, exergy, and pollution, considered as independent criterions. The evident target of every engineering project, especially of energy ones, is to have a minimum of the above mentioned expenditures. However, a reduction of one of them often leads to the increase of another one. It is just the case of Pareto optimization.

A simple model to connect all the criteria to exergy efficiency is used. The increase of exergy efficiency above the optimal value gives a decline of the current part of expenditures, but increases the invested part of them, needed to manufacture equipment. The approaching of exergy efficiency to unity leads to the infinite size of equipment along with prohibitive expenditure of exergy and money, as well as pollution release.

The three-dimensional characteristic curve is depicted in a rectangular coordinate frame with an indication of the three individual minimum values.

The regular procedure of Pareto optimization is demonstrated by the use of three priority coefficients. The Pareto-optimal exergy efficiency is presented in a simple analytical form. The careful selection of the three numeric correlation parameters for the dependence of the invested exergy, capital, and associated pollution upon the exergy efficiency is needed.

Optimization is the most common problem of every engineering activity. Actually, a design is just an optimization in the frame of restrictions of the real world. One of the best illustrations is the book of Bejan et al. (1996), where the words "design" and "optimization" in the title are in a neighborhood. An extensive bibliography might be found there.

A remarkable event was the summer school of NATO Institute of Advanced Studies in July 1998 in Neptun, Romania (Bejan and Mamut, 1999).

As a rule, optimization problems are based on complicated models, where only computer simulations might be used and an analytical solution is impossible. Generally, the target function is only one, cost of energy (COE) or entropy gain or exergy destruction minimization. The very important consideration on resources scarcity or pollution mitigation are taken into account by monetary cost of exergy (Bejan et al., 1996) and pollution with subsequent minimization of overall costs. Real life, however, sometimes forces the use of many independent target functions simultaneously, when a decline of one of them causes the increase of another one. How to find the optimal solution as a base for decision making? Here is just the place for Wilfredo Pareto's approach (von Neumann and Morgenstern, 1947; Steuer, 1986).

In energy projects, the most important is minimization of the three target functions: exergy, monetary expenditure, and pollution release, which are considered quite independent.

Presentation of these three target functions in the form of a three-dimensional "space of social will" has been proposed in Yantovsky (1994, p. 62): "Many values might be taken into account to describe energy supply."

According to the thoughts of Yantovsky, the best are three: "monetary cost of energy, sum of specific exergy consumption, pollution or risk to mankind... The palliative solution to this problem might be found if we consider all three values as quite independent and award each item by its own coordinate axis… Any decision maker wants to launch an optimal issue. Here, well-known Pareto optimization might be used…"

At that time, however, there were no analytical models to deal with these quantities. Now, we have the branch of exergy analysis referred to as "exergonomics," which considers not only current exergy expenditure, but also the former one, invested to produce equipment (Yantovsky, 1994). It is similar to Exergy Life Cycle Analysis, described in detail by Cornelessen (1997) and Gong and Wall (1997). Both of these works give background to our optimization problem. We prefer the name "exergonomics" to stress the similarity to ordinary economics.

The aim of the present section is to demonstrate the Pareto-optimization procedure in the three target function frame ("energy trilemma") on the basis of a simple model of the link of exergy expenditure, monetary cost, and pollution with exergy efficiency of an energy object (power plant, boiler house, refinery, etc.).

We would like to clearly indicate the difference of our approach from well-known advanced approaches to the problem of power systems optimization (Makarov et al., 1998; Song, 1999).

We consider not existing, but only future energy systems in the time of its design; therefore, all the invested expenditures are of importance.

We consider not only power, but also thermal and chemical energy consumption and delivery, evaluated in exergy units.

We simultaneously consider the three most important target functions. They are visualized in a three-dimensional rectangular frame system, entitled *social will system*, because serving as a society designer tends to minimize all the three.

The need to make some changes in the external conditions of an energy system operation, which would introduce uncertainty and additional expenditures for adaptation (Makarov et al., 1998) is beyond the scope of this chapter. Recent findings for modern optimization techniques (Song, 1999), such as simulated annealing, genetic algorithm, neural network, fuzzy logic, Lagrangial relaxation, and ant colony might be some subjects for future works on Pareto optimization.

9.5.2 INVESTED AND CURRENT EXPENDITURES

Considering exergy, monetary, and pollution expenditures, we will distinguish the invested part and the current part. Speaking on pollution expenditure, we mean the flow of a pollution substance, such as combustion products, particulate matter, etc., which should be minimized.

The *invested* part of every expenditure is what we need to produce equipment and construct the buildings. It means the capital investment, the exergy consumed to extract raw materials, to melt metals, to roll, cut, forge, and produce construction materials, transportation, etc., and pollution released during the mentioned activity.

The *current* (or operational) part is what we spend in normal operation, such as the fuel consumption and associated payments, fuel exergy flow, and combustion products such as pollution flow. The *delivered* exergy flow is the reason for the considered object existence, exergy of delivered electrical power, thermal energy in district heating, or oil products from a refinery. All mentioned exergy parts are presented in Figure 9.9 versus time. Note the main optimizing parameter, exergy efficiency, as the ratio of delivered to consumed current (operational) exergy flow. Note also that, in Figure 9.9, the invested exergy ε_b is the amount measured in joules (subscript "b" means "building"), whereas delivered and consumed are exergy flows, measured in watts.

There exists the link between the invested and current parts of any expenditure. We always can decrease the boiler fuel consumption by decreasing the temperature drop; however, this action will increase the heat-transfer surface, and hence, the size and exergy consumption on the metallurgical mill to produce metal, which is the invested part of expenditures. The Carnot ideal engine with exergy efficiency near unity is impossible to construct due to the limitless invested exergy expenditure. These physical reasons suggest that there exists an optimal relation between invested and current expenditures. However, there are no reasons that they coincide for the three targets. Actually, they are different. The principal problem of optimization is to find a simple enough analytical model to describe the link between the invested and current parts of expenditures.

9.5.3 EXERGY MINIMIZATION

Let us first consider the target function for exergy. The simple model, connected exergy efficiency with invested exergy (Iantovski, 1998) is

$$\varepsilon_b = \varepsilon_d \cdot \tau / (1 - \eta)c^2 \tag{9.87}$$

Here, invested exergy is inversely proportional to the exergy destruction rate. The correlation parameter c is used to adjust to known numerical data of equipment.

The total exergy efficiency, taking into account the sum of invested and current expenditures equals

$$\eta_{\text{tot}} = \varepsilon_d \cdot \tau / (\varepsilon_{\text{con}} \cdot \tau + \varepsilon_b) \tag{9.88}$$

The main criterion in exergonomics, the specific exergy consumption is:

$$Z = 1/\eta_{\text{tot}} = 1/\eta + 1/K, \quad \eta = \varepsilon_d / \varepsilon_{\text{con}}, \quad K = \varepsilon_d \tau / \varepsilon_{\text{con}} \tag{9.89}$$

Introducing Equation (9.88) into K, we get

$$K = c^2 (1 - \eta); \quad Z = 1/\eta + 1/(1 - \eta)c^2 \tag{9.90}$$

From $dZ/d\eta = 0$, we get an individual optimum

$$\eta_{\text{opt}} = c/(c+1); \quad Z_{\min} = 1/\eta_{\text{opt}}^2 \tag{9.91}$$

The geometric locus of maximal η_{tot} is the square parabola; see Figure 9.15 a, b. Here, the three curves reflect different (increasing) correlation parameter c cases. The more c, the more optimal exergy efficiency according to Equation (9.91).

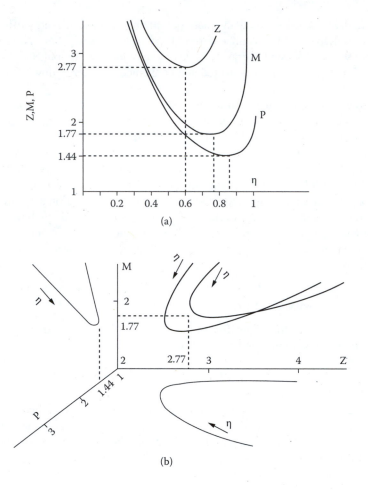

FIGURE 9.15 (a) Dimension criterions Z, M, P, versus exergy efficiency; (b) characteristic curve in the 3-D frame Z-M-P (exergy—money—pollution) and its projections (Yantovsky, 1994); c = 1.5, m = 3, p = 5.

9.5.4 MONETARY AND POLLUTION OPTIMIZATION

The monetary cost of a product is the specific cost of delivered exergy unit multiplied by exergy flow rate and lifetime (Bejan et al., 1996):

$$\lambda_d \cdot \varepsilon \cdot d \cdot \tau = \lambda_{con} \cdot \varepsilon_{con} \cdot \tau + \lambda_b \cdot \varepsilon_b \tag{9.92}$$

Introducing the correlation for exergy according to Equation (9.92), we get:

$$\lambda_d = \lambda_{con}[1/\eta + \lambda_b/c^2/(1-\eta)] \tag{9.93}$$

It is evident that optima for monetary cost and exergy should be different, because costing changes the correlation parameter. If we introduce the new parameter

$m = c\sqrt{\lambda_{con}/\lambda_b}$, all the previous results are the same by replacing c for m:

$$M = \lambda_d/\lambda_{con} = 1/\eta + 1/(1-\eta)m^2 \qquad (9.94)$$

The total pollution release, associated with delivered exergy is:

$$r_d \cdot \varepsilon_d \cdot \tau = r_{con} \cdot \varepsilon_{con} \cdot \tau + r_b \cdot \varepsilon_b \qquad (9.95)$$

The pollution from fuel-fired power plants consists of different exhaust gases with different environmental impact. In order to reduce all the gases to only one equivalent flow rate, the environmental impact units (EIUs) are used (Muessig, 1999). One EIU equals 1 t of carbon dioxide (CO_2) emission. In this unit, the emission of 1 t $[SO_2]$ = 93 EIU, 1 t $[NO_x]$ = 172 EIU, and 1 t $[CH_4]$ = 29 EIU. The total pollution is the sum of products of gas mass fractions and its EIUs.

Introducing a correlation parameter for pollution $p^2 = c^2 + r_{con}/r_b$, we get:

$$P = r_d/r_{con} = 1/\eta + 1/(1-\eta)p^2 \qquad (9.96)$$

It is the dimensionless pollution criterion, similar to Z for exergy. Equations (9.90), (9.94), and (9.96) describe the characteristic curve in the three-dimensional frame Z-M-P as functions of one optimized parameter η.

9.5.5 PARETO OPTIMIZATION PROCEDURE

Let us denote as η_z the individually optimal exergy efficiency for Z criterion, and accordingly η_m for the M criterion and η_p for P criterion.

As a desired compromise, let us find such a solution η^* which gives minimal deviation from every optimal value:

$$F(\eta^*) = \min\{u \cdot [Z(\eta) - Z(\eta_z)] + v \cdot [M(\eta) - M(\eta_m)] + w \cdot [P(\eta) - P(\eta_p)]\} \qquad (9.97)$$

Where u, v, w, are some weight (priority) coefficients by condition

$$u + v + w = 1 \qquad (9.98)$$

One of many ways to evaluate the priority coefficients is polling of an expert group or shareholders as to what is for them the most important: resources (exergy), cost, or pollution minimization. The number of votes "for" is to be divided by the number of all voters.

As it was first shown by Zack (1972), all the values of η^* as solutions of Equation (9.97) form the Pareto set by different coefficients u, v, w.

We see that

$$u \cdot Z(\eta_z) + v \cdot M(\eta_m) + w \cdot P(\eta_p) = const,$$

and the search of our desired compromise solution is the minimization of the functional

$$F(\eta) = 1/\eta + u/(1-\eta)c^2 + v/(1-\eta)m^2 + w/(1-\eta)p^2 \qquad (9.99)$$

From Equation (9.99), we get

$$\partial F/\partial \eta = -1/\eta^2 + 1/(1-\eta)A^2 = 0,$$

where

$$A^2 = 1/(u/c^2 + v/m^2 + w/p^2) \qquad (9.100)$$

The coincidence to the case of Equation (9.100) and Equation (9.101) is evident:

$$\eta^* = A/(A+1) \qquad (9.101)$$

If one assumes $v = w = 0$, $A = c$, then $\eta^* = \eta_z$ and the like for the other two priority coefficients.

Changing coefficients u, v, w by given c, m, p, we get the Pareto set. Let us go to some visualization of described results.

9.5.6 NUMERIC ILLUSTRATIONS

Suppose the numeric values of correlation coefficient (from the statistics on exergy expenditures, costs, and specific pollution release) are $c = 1.5$, $m = 3$, $p = 5$. According to formula (9.91), valid by arbitrary c, we can calculate the individual (not Pareto) optima and minima:

TABLE 9.5
Optima for Selected Correlation Parameters

Parameter		η_{opt}	Minima
c	1.5	0.60	Z = 2.77
m	3	0.75	M = 1.77
p	5	0.833	P = 1.44

Source: Yantovsky, E. and Zack, Y., 2000.

The curves for Z, M, P versus exergy efficiency are presented on Figure 9.16. Their increase after minimal value reflects the increase of invested exergy, when exergy efficiency approaches unity. All three optimal η and three minima are indicated. The arrow shows the direction of exergy efficiency increase. If we consider a heat exchanger, the optimizing parameter might be the temperature drop in a wall. In this case, the arrow direction is opposite.

TABLE 9.6
Example of Pareto Optimization Results

		A	η^*	Z	M	P
c = 1.5	u = 0.1	3.135	0.785	3.155	1.77	1.484
m = 3	v = 0.3					
p = 5	w = 0.6	**1.165**	**0.538**	**10.516**	**4.022**	**1.993**
	u = 0.6	1.860	0.650	2.808	1.855	1.652
	v = 0.1					
	w = 0.3	**0.630**	**0.386**	**9.10**	**4.22**	**2.691**
c = 0.5	*u = 0.3*	*2.214*	*0.688*	*2.877*	*1.809*	*1.581*
m = 1	*v = 0.6*					
p = 4	*w = 0.1*	*0.745*	*0.427*	*9.321*	*4.087*	*2.450*

Source: Yantovsky, E., 2000.

The characteristic curve in the rectangular frame of social will and its projections are presented in Figure 9.11. The three individual optima are indicated.

Some numeric results of the Pareto-optimization are presented in Table 9.6 for the two groups of given correlation parameters c, m and p and three values of priority coefficients for each group. When priority of exergy is increased from 0.1 to 0.6, the exergy expenditure criterion Z declines from 3.155 to 2.808 or from 10.516 to 9.10. *The second group of c, m, p and associated numbers are printed in italic.*

The table shows a sensitivity of the Pareto-optimal exergy efficiency to the values of priority coefficients u, v, w, sufficient for a robust determination of optima. In an interesting paper on heat-exchanger design, D. Sama (1993) felt sorry for telling students

> ... that they must design only at the exact economic optimum. To do otherwise would be heretical, thus we send them on in search of the holy grail called the 'global optimum'... The existence of a global optimum is a myth whose pursuit is fruitless.

We agree with this statement only if a designer does not know what he actually wants. If he properly selected not only one, but an arbitrary couple of target functions and specified the priority coefficients, he can get the global optimum. It is just the Pareto one.

To create the simple model of the link between the invested and delivered exergy, let us find an analytical solution for Pareto-optimization of energy objects in the three-dimensional space (exergy-money-pollution). The three correlation parameters should be taken by means of statistical data. This approach is worth the attention the decision makers in energy engineering and policy.

REFERENCES

Arons, G. 1926. "On cost estimation of electrical energy and steam in combined thermal power plants," in *Proc. of the 3rd All-Union Congress for Heat Engineering*, Moscow, Nov.10–18, Issues of Heat Engineering Inst, Moscow, 1927, v.3, Iss.1: 109–117 (in Russian).

Bejan, A. and E. Mamut (eds.). 1999. *Thermodynamic Optimization of Complex Energy Systems*. London: Kluver, NATO Sci. Ser.

Bejan, A. et al. (eds.). 1998. *Proc. of Efficiency, Cost, Simulation and Environmental Aspects of Energy Systems and Processes ECOS'98*, July 8–10, Nancy, France.

Bejan, A., Tsatsaronis, G., and M. Moran. 1996. *Thermal Design and Optimization*. New York: John Wiley & Sons, Inc.

Bishop, D., Giles, C.R., and S. Das. 2001. "The rise of optical switching," *Sci. Am.* 284(1): 88–94.

Bolland, O. and I. Ertesvag. 2001. *Exergy Analysis of Gas-Turbine Combined Cycle with CO_2 Capture using Auto-Thermal Reforming of Natural Gas*, www.ept.ntnu.no/ noco2/ poster%20exergy%20Liege202001.ppt (accessed July 2, 2008).

Chenery, H.B. 1949. Engineering production functions. *Quart. Jour. of Economics*, (63): 507–531.

Corneliessen, R.L. 1997. Thermodynamics and Sustainable Development. Ph.D. thesis. University of Twente, Enschede, the Netherlands.

De Groot, S. and P. Mazur. 1983. *Non-Equilibrium Thermodynamics*. New York: Dover. Iantovski, E. 1997a. "Exergy and Information Currents," in *Proc. of Florence World Energy Res. Symp. FLOWERS'97*, July 30, Italy, SG Editoreali, Padova, 921–929.

De Vries, B. and E. Nieuwlaar. 1981. A dynamic cost-exergy evaluation of steam and power generation. *Resources and Energy*, No. 3: 359–388.

Gaggioli, R. and Y.A. El-Sayed. 1987. "Critical Review of Second Law Costing Methods," in *Proc. 4th Int Symp. Second Law Analysis of Thermal Systems*, Moran, M. and E. Sciubba (Eds.), May 25–27, Rome, Italy, 59–73.

Gas Technology Institute. 2004. *A Technology Development Prospectus. High Efficiency; Ultra-Low Emissions, Thermochemically Recuperated Reciprocating Internal Combustion Engine Systems*.

Gibbs, J.W. 1928. *Graphical Methods in the Thermodynamics of Fluids. The Collected Works of J.W. Gibbs*, New York: Longmans Green.

Gong, M. and G. Wall. 1997. "On Exergetics, Economics and Optimization of Technical Processes to Meet Environmental Conditions," in *Proc. Int. Conf. TAIES'97*, Beijing, June 10–13, 453–460.

Hoverton, M.T. 1964. *Engineering Thermodynamics*. New York: Van Nostrand, Princeton.

Iantovski, E. 1997a. "What is Exergonomics," in *Proc. of Florence World Energy Res. Symp. FLOWERS '97*, July 30, Florence, Italy, 1163–1176.

Iantovski, E. 1997b. Exergy current vector in Energy Engineering, in *Proc. Int. Conf. Thermodynamic Analysis and Improvement of Energy Systems*, TAIES'97, June 10–13, Beijing, China, 641–647.

Iantovski, E. 1998. "Exergonomic optimization of temperature drop in heat transfer through a wall", in *Proc. of Efficiency, Cost, Simulation and Environmental Aspects of Energy Systems and Processes ECOS'98*, 8-10 July, Nancy, France, pp. 339–343.

Iantovski, E. and W. Seifriz. 1996. "An Attempt to Include Information Flow into Exergy Current," in *Proc. ASME Int. Conf. on Engng Syst. Design & Anal, ESDA'96*, July 4, Montpellier, France, vol.73, 215–220.

Ishida, M. et al. (eds.) 1999. *Proc. of Int. Conf. ECOS'99*, June 30, Tokyo, Japan.

Khazen, A. 1991. Peculiarities of the Second Principle of Thermodynamics Application for the description of the Brain Work, *Biophysics*, 36(4): 717–728.

Kreuzer, H.J. 1984. *Nonequilibrium Thermodynamics*. Oxford: Clarendon Press.

Landau, L. and E. Lifshits. 1982. *Electrodynamics of Continua*, Moscow: Izd. Nauka.

Leff, H.S. and A.F. Rex (eds.). 1990. *Maxwell's Demon, Entropy, Information, Computing*. Princeton: Princeton Univ. Press.

Lieb, E. and J. Yngvason. 2000. "A fresh look at entropy and the second law of thermodynamics." *Phys. Today*, (53): 32–37.

Lightman, A. 1992. *Great Ideas in Physics*. New York: McGraw-Hill.

Machta, J. 1999. Entropy, information and computation. *Am. J. Phys.* 67(12): 1074–1077.

Makarov, A.A. et al., 1998. "Decision-making Procedure for Optimizing Power sources," *Applied Energy: Russian Jour.* Vol. 36, No. 5 (in Russian).

Muessig, S. 1999. "New Technologies and Methods," *Oil/Gas Eur. Magazine*, No. 4, 14–18.

Proceedings of Int. Conf. 1999. *Efficiency, Cost, Optimization, Simulation and Environmental aspects of Energy Systems,* Ishida, M., et al. (eds.), June 8–10, Tokyo, Japan.

Reif, F. 1999. Thermal Physics in the introductory course: Why and how to teach it from a unified atomic perspective. *Am. J. Phys.*, 67 (12): 1051–1062.

Rivero, R. et al. (Ed.) 2004. *Energy-Efficient, Cost-Effective, and Environmentally-Sustainable Systems and Processes*. Mexico: Instituto Mexicano del Petroleo. pp. 801–817.

Sama, D. 1993. "The use of the Second Law of Thermodynamics in the design of heat exchangers," in *Proc. Int. Conf. Energy Systems and Ecology*, Cracow, Poland, July 5–9, 123–132.

Schlögl, F. 1989. *Probability and Heat: Fundamentals of Thermostatistics*. Braunschweig: Vieweg.

Slepian, J. 1942. *Energy Flow in Electrical Systems Trans. AJEE*, vol. 61, 835–840.

Smil, V. 1991. *General Energetics*. New York: J. Wiley & Sons, Inc.

Soddy, F. 1926. *Wealth, Virtual Wealth and Debts*. London: G. Allen and Unwin.

Song, Yong-Hua (Ed.) 1999. *Modern Optimisation Techniques in Power Systems*. Dorderecht: Kluwer Acad. Publ.

Steuer, R.E. 1986. *Multiple Criteria Optimization: Theory, Computation and Application*. New York: J. Wiley & Sons, Inc., 546.

Sutton, G. and A. Sherman. 1965. *Engineering Magnetohydrodynamics*. New York: McGraw-Hill.

Szargut, J. 1971. "Anwendung der Exergie zur angenaherten wirtschaftlichen Optimierung," *Brennstoff, Wärme, Kraft* (BWK), (23):516–519 (in German).

Szargut, J. 2005. *Exergy method. Technical and Ecological Applications*. New York: WIT Press.

Szargut, J. and D. Morris. 1987. Cumulative Exergy Consumption. *Energy Research*, (11): 245–261.

Szargut, J. and R. Petela. 1965. *Egzergia*. Warszawa: Wyd. Naukowo-Techniczne.

Tsatsaronis, G. 1987. "A review of exergoeconomic methodologies," in *Proc. 4th Intern. Symp. Second Law Analysis of Thermal Systems*, Rome, May 25–27, 81–86.

von Neumann, J. and O. Morgenstern. 1947. *Theory of Games and Economic Behaviour*. Princeton: Princeton Univ. Press.

Wall, G. 1977. "*Exergy — a useful concept within resource accounting,*" Report No.77-42. Chalmers University, Göteborg, Norway.

Yantovsky, E. 1984. "A method of thermodynamic effectiveness calculation by the sum of specific exergy consumption." In *Energetics and Fuel*. Int. Centre of Sci. and Techn. Information, Acad. Sci. USSR, Moscow, No.6, 82–94 (in Russian).

Yantovsky, E. 1989. Non-equilibrium Thermodynamics in Thermal Engineering. *Energy: The Int. Jour.* 14(7): 393–396.

Yantovsky, E. 1994. *Energy and Exergy Currents*. New York: NOVA Sci. Publ.

Yantovsky, E. and J. Poustovalov. 1982. *Compressional Heat Pumps*. Moscow: Izd. Energoizdat (in Russian).

Yantovsky, E. et al. 2004. "A Zero Emission Membrane Piston Engine System (ZEMPES) for a bus," presented at Int. Conf. *Vehicles Alternative Fuel Systems and Environmental Protection* (VAFSEP), July 6–9, Dublin, Ireland, 129–133.

Yantovsky, E. and Y. Zack. 2000. "Pareto Optimization in Exergonomics," *Proc. ECOS-2000*, The Netherlands, July, 1, 343–356.

Zack, Y. 1972. "Models and methods of compromise planning in mathematical programming with many target functions," *Cybernetics*, No. 4, 102–108 (in Russian).

Zemanski, M. 1968. *Heat and Thermodynamics*. 5th ed., New York: McGraw-Hill.

10 Two Lectures for Students and Faculty of Dublin Institute of Technology (2003)

10.1 TO BAN OR NOT TO BAN?
(ON THE HUMAN RIGHT TO BREATHE
AND GLOBAL WARMING)

Ancient citizens of European cities had an unpleasant custom of throwing the contents of night pots away through windows. After the ban of this practice, many people continued throwing nightly over the fences to neighbors. Only at the end of the nineteenth century, were sewage systems developed to cope with the problem of biological human wastes.

Technological wastes (including power plants and boiler houses) exceed by mass the biological ones. The main part of technological wastes is exhaust gases discharged into the atmosphere after combustion of different fuels. Look at the roofs in old European cities. You will see thousands of stacks or chimneys. Isn't it the same as throwing the waste through windows? When a source of smog is too large, a very high stack of more than 300 m is erected in the vicinity to throw wastes to more distant neighbors.

Now imagine you are living on a river. Nobody may force you to drink dirty water, polluted by a landlord upstream. He will be punished for polluting. That is why recently European rivers have become cleaner, even with nearby fisheries. Why, then, should you breathe dirty air, polluted by power plants, heating systems, industry, or vehicles? Why is it impossible now to completely forbid all the discharges into the atmosphere? You and your children are working in this industry and drive these vehicles.

Recently, many manufacturers have shifted far away from Europe, but Europeans might be afflicted due to loss of jobs. In general, such a shift does not differ from throwing away waste at the neighbors as in ancient times. If the neighbors are hungry, the possibility to earn overtakes the fear of pollution. This is the case of the developing countries, including Eastern Europe. But this agreement is temporal. More drastic measures are needed. The problem of technological wastes might be further ignored by decision makers, leaving it as merely slogans of "green" parties. But in recent decades, it has attracted new attention due to the menace of global warming or climate change.

The additional heating of the atmosphere was forecasted in 1827 by Joseph Fourier, a French physicist and mathematician, known as the "Newton of heat transfer."

In 1896, one of the founders of physical chemistry, Swante Arrhenius, described the nature of the phenomenon in the paper, *On the Influence of Carbonic Acid in the Air upon the Temperature of the Ground*. Solar beams carry a lion's share of energy by the short-wave ultraviolet flow when they are falling to Earth. All the incidental energy should be re-radiated back to space. But the temperature of the ground is low and beams are long-wave, infrared. Existing in the atmosphere, carbon dioxide and some other trace gases absorb the waves of just this length, forming a blanket around the planet. In principle, this phenomenon is extremely important; otherwise, the ground would be too cool, and without life. The glass in a greenhouse plays a similar role of a blanket. That is why the effect is called a "greenhouse effect." It is a normally existing effect; harmful it is increased due to manmade gases. On the planet Venus, the atmosphere consists mostly of carbon dioxide, its surface temperature is about 500°C and thus no life is possible.

The works by Fourier and Arrhenius have been ignored until the late 1970s when a leading climatologist in the Soviet Union, M. Budyko, had compared their views and his own, showing the reality of the menace in the book *Carbon Dioxide in Atmosphere and Climate* in 1973. The entire world began researching correlations of carbon dioxide concentrations and surface temperature, measuring it at thousands of observatories on land, by planes, and satellites. Powerful computers using sophisticated models have been used to predict climate changes.

The main international scientific body, IPCC (Intergovernmental Panel on Climate Change), headed by Professor B. Bolin stated in 1993:

> We are certain that emissions from human activity are substantially enhancing the concentrations of greenhouse gases in the atmosphere … These increases will result in an additional warming of the Earth surface. We calculated with confidence that … fossil fuel use has been responsible for more than half of the ongoing change of the enhanced greenhouse effect. Long lived gases would require immediate reductions of emissions from human activities by more than 60% to stabilize their concentrations at today's levels.

In 1996, the same IPCC added to this the possibility of a more dangerous hydrological cycle, affecting such extreme events as droughts, floods, and rainfalls. The unexpected flood in Central Europe is being treated by many as a confirmation of the IPCC forecast.

Summits of world leaders in Rio de Janeiro, Morocco, and Johannesburg were devoted to prevention of global warming. In 1997, a protocol was issued in Kyoto about damping of greenhouse gas emissions. It has already been signed by 120 countries, but in some it is not yet ratified.

So far, so good, but not clear. Many climatologists flatly reject the explanation of the phenomenon and even the existence of global warming. Some of them have stated that emissions of methane from swamps and cracks in the Earth's crust far exceed the manmade release. Even the total halt of this release will have little effect with respect to natural emissions.

Other experts said that observed warming does not have any link to gas concentration. They opined that in the geological history of the Earth, the carbon dioxide concentration in the atmosphere was ten times more than in the contemporary period but that it caused only the ice period to occur, not warming. The independent measurements of the water temperature in the North Ocean, made by enemies during 40 years of the Cold War revealed a decrease, not an increase in temperature. Some experts stated that the satellite measurements have statistically proven that warming has not been observed at all.

The controversy sometimes is so sharp, that it spoils any truth finding. An example is as follows.

The Russian Institute of Global Climate and Ecology (RIGCE), headed by academician Yu Israel stated, "General trend still remained unchanged, there exists climatic warming in Russia and the world as well." But academician K. Kondratiev when interviewed on March 29, 2004 stated:

> To mitigate emissions into atmosphere is not realistic even if it is needed ... The RIGCE says nonsense. In the scientific world of climate problem exists a powerful mafia ... thousands of people are corrupted in order to write in support of some concepts. And I am not a single dissident.

There are enough of nay-sayers on the global warming theory in the world. Note the definite and strong sentences.

However, during this same time period, an equally strong opposite opinion was expressed by Professor David King, the chief consultant for the British Government in the magazine *Science*, on January 9, 2004. He stated that the climate change menace is more harmful for mankind than terrorism. He reproached the U.S.A. for neglecting to do anything against global warming. He maintains that U.S. emissions are equal to 20% of the world, when the U.S. population is only 4%.

The damage to the U.K. he estimated as tens of billion of pounds annually due to the sea level rise from water warming and melting near Greenland and Antarctic ice. He forecasts a rise of storm waves with increased coastal destruction. Urging reduction of carbon dioxide emission, he argued that a delay of urgent measures will cause more expensive actions later. He estimated the contemporary spending as 1% of the gross domestic product (GDP) of developed countries.

In these short notices, it is impossible to describe the climate controversies on a full scale. Most dangerous is the lengthy time of the discussions. The only judge in natural sciences, His Majesty Experiment in global scale can be made by Lord only whereas all the experiments on small models will not assure. In this setting, there appears an intention to find a background of preventing measures from the point of our beginning. To rationalize the human right to expel emissions into the atmosphere, would be equal to permitting pollution of a river or even to the throwing of wastes through windows. As usual, when one tries to show the importance of something for a man, one will always indicate the time span of living without it. So, without food — months, without water —weeks, without breathing — only minutes. Do not minimize the problem by stating that "without breathing" is still not the case and even in large cities, there is enough fresh air. It is important to stop the trend before it is too late. Every discharge into the atmosphere is the violation of the human right to breathe. The robbery of one person is considered a heavy crime regardless of the fact that there is a large amount of other people who have not been robbed.

The immediate question comes to mind of whether it is possible to ban every discharge of waste into the atmosphere.

Most frequently, it is answered that we should stop fuel combustion and turn to an energy supply without combustion. Such a supply involves nuclear power and solar-, wind-, and hydropower, collected under the title of renewable energy. There exists a firm negative attitude to nuclear power due to still unsolved problems of nuclear wastes and as a secondary nuclear weapon for terrorists, especially after the event of September 11, 2001.

Renewables might form the basis of an energy supply in some distant future, but they are unable to dominate in this century. There are no alternatives to fossil fuels for power plants and vehicles for at least half a century. But this doesn't mean that emissions are unavoidable.

The technical solution is simple: to convert gaseous wastes into liquid ones (emissions into effluents) and then to follow the treatment of biological wastes, the sewage system. The main product of combustion, carbon dioxide in liquid form, has a thousand times less volume, and might be transported under pressure of about 100 bars by a steel pipe system and then sequestered deep underground or in the ocean's depth. Such a solution was first offered by an Italian engineer, Cesare Marchetti, in 1977. The scheme of a fuel-fired power unit, cogenerating power and liquid carbon dioxide has been patented in the Soviet Union by V.L. Degtiarev and V.P. Gribovski since 1967. But they did not know of the greenhouse effect and developed the unit for industry.

In cooperation with colleagues from different countries, this author has been concerned for 15 years with the development of a fuel-fired power plant without exhaust gases. The scheme is comprised of an air separation unit to produce oxygen and release harmless cold nitrogen back into the atmosphere. Combustion takes place in "artificial air," a mixture of oxygen and recirculated carbon dioxide. The use of this artificial air is as a working substance in turbines and deflection of a fraction of it from the cycle to transport for sequestration. The fraction is exactly equal to an amount of carbon dioxide formation in the combustion chamber. Such a scheme was first described in 1991 at the World Clean Energy Conference in Geneva. The most recent version with the detailed references list could be easily found at: www.carbonsq.com/pdf/ posters/CSII5.pdf, and in application to piston engines for vehicles at: www.carbonsq.com/ pdf/posters/CAPI1.pdf.

In 1991, The International Energy Agency created the Greenhouse Gas R&D Programme in Cheltenham, U.K. under the direction of a group of scientists to coordinate worldwide research and publish a bimonthly magazine called *Greenhouse Issues*.

During the last decade, combustion without emissions has been discussed at six International Meetings On Greenhouse Gas Control Technologies, three National Conferences in the U.S.A. on carbon sequestration and many other meetings. In particular, at Liege University (Belgium) in 1997 and 1998 were held Workshops on Zero Emissions Power Cycles, where it was shown that every scheme of a fuel-fired power unit might be converted into a zero emissions version almost without sacrificing efficiency. The state-of-the-art of zero emissions technology is well described in www.iea.org/impagr/zets/status/2002tsr.pdf. At a recent Conference VAFSEP in Dublin on July 6–9, 2004, the paper "Zero Emission Membrane Piston Engine system (ZEMPES) for a Bus" was presented by my colleagues Dr. James McGovern, Professors Shokotov and V. Vaddella. It might be a good subject for further elaboration in DIT.

The fate of sequestered carbon dioxide was the subject of investigation by many geologists. The most important result is the successful experiment in Norway, where one million tons of carbon dioxide were injected annually to the depth of 800 m over a period of 8 years. Still no hints of escaping or other problems are observed. Carbon dioxide eventually is bound in the form of calcium carbonate (school days chalk for blackboard). The amount of strong brines deep underground (hydro lithosphere) is so large that all the world technological wastes might be sequestered for a millennia. It is possible to state that, for the protection of the human right to breathe, all the exhaust gases should be directed not into the atmosphere but into the hydro lithosphere. The same action would protect the Earth from global warming due to the greenhouse effect, if it really exists.

What is the setting in different countries?

In Germany, the climate change problem attracted great attention. The Green Party takes more and more places in Bundestag. Its member, Jurgen Trittin, as the Minister of Environment Protection and Reactors, has made many important decisions. At the International Conference in Bonn on July 1, 2004, he confirmed the gradual shutting down of nuclear reactors with the decommissioning of the last in about 20 years. Germany is adhering well to her obligations according to the Kyoto protocol. By 2008, the emissions would be decreased by 21% with respect to 1990. In 2000, the actual decrease was 19.1% which is a good example for all the developed countries.

The concept of zero emissions power has been examined by a major group of specialists from universities and industry entitled COORETEC. Their report in December 2003 contained, in particular, the perspective of ceramic membranes for oxygen production for zero emissions power plants. They set oxygen apart from nitrogen and other gases in the air by temperatures of about 900°C. This forms the mixture of oxygen and carbon dioxide on a much cheaper and simpler scale than ordinary cryogenic air separation units. Unfortunately, it is unknown whether in Germany a demonstration unit is underway. In the United States, such a unit soon will be built by the company, Clean Energy Systems, in California, (www.cleanenergysystems.com).

The CES Company has been arranged by a group of retired German co-workers of Werner von Braun in creation of big rockets. They know well how to combust in oxygen. The main idea and patent of 1996 belongs to Rudolf Beichel (1913–1999). His scheme does not differ from the one presented at the Geneva World Clean Energy Conference in 1991.

Very active against emissions is former British Prime Minister Tony Blair. He shared opinions of David King and created an International Climate Group for Emissions mitigation, which will be active irrelevant to the process of sign and ratification of Kyoto protocol. Tony Blair promised next year, when the U.K. is to be the chair in the "Big 8," these questions would dominate in their agenda. The U.K. had demonstrated that damping emissions is not detrimental to economic growth, which had risen in 1990 to 2002 by 30% along with the emissions decline by 15%. He states that the U.K. will damp emissions by 60% in 2050.

Very interesting is the U.S. administration policy. On one side, they flatly rejected the protocol on emissions damping. The United States still dominates the nay-sayers on global warming harm. From the other side, all the research and development works toward zero emissions are generously supported by the Department of Energy and their results are actively discussed at regular meetings of high scientific rank. Many interesting works are underway in Japan, especially in the area of conversion of carbon dioxide into transportation fuels. Powerful industrial companies joined their efforts in carbon dioxide capture and sequestration; see www.co2captureproject.com.

Finally, some words on the recent events in Russia, which were not in agreement with the Kyoto protocol. Last summer (2002)*, after the Climate Conference in Moscow, President Putin asked Russian scientists to analyze the reason to ratify the Kyoto protocol. The Academy of Sciences formed a special council, which held 12 sessions with 20 presentations by leading academicians. Unfortunately, the opinions were different as it was shown in controversy by Izrael-Kondratiev. To the last session of this council were invited many foreign climatologists and other specialists. A great scandal unknown in the scientific world occurred at this session. Andrey Illarionov, adviser to the President, and a known enemy of the protocol, had compared the global warming theory with a totalitarian ideology such as Nazism, Marxism, or Lysenkoism. Europe was already under the power of the "brown pest," said Illarionov. Now the pest of another color is coming and again Russia stands on its way.

A fraction of indignant foreign participants headed by David King could not tolerate this speech and left the session. The comparison is not only notorious but symptomatic. At first, Russia had crucially supported these theories, and implemented their terrible practice, and only then after implementation was against them. But for Russia, there would not have been the split between communists and social-democrats, which paved the legal way to Nazism power. But for Russia, Marxism would have had only modest places in economic textbooks along with the other theories.

The absurdity of the Lysenko theory was evident to all the honest biologists in Russia, but the cads in power theoretically killed the national Russian genius, academician Nicolay Vavilov and annihilated Russian genetics.

Illarionov's comparison and his administrative position lead to a natural question: If your struggle against the Kyoto protocol would end in the same manner as that of your predecessors against genetics, who would be responsible for that?

Being unaware of an answer to this question and being acutely aware of the existence of global warming, I will risk a hope: at some time in the future, all the emissions into the atmosphere would be forbidden as a criminal violation of the human right to breathe. Zero emissions technology is unavoidable.

10.2 ON THE FATE OF A MECHANICAL ENGINEER (LECTURE AT DUBLIN INSTITUTE OF TECHNOLOGY, OCTOBER 30, 2003) BY PROFESSOR DR. E. YANTOVSKY

When I was in Ireland for the first time, the income per capita in the country was modest, less than in major European countries. Now in 2003, I'm privileged to speak in the richest country of Europe. It does not mean that you are especially industrious. Russians are industrious, too. It means you have a clever government.

---

* For update, see modern political statements in Preface and Concluding Remarks.

One of discoverers of social laws (perhaps, Nortcott Parkinson, but I'm not sure) stated: the stupidity of a government is directly proportional to the size of the country. Accordingly, when a country is small, the government is clever.

LECTURER OR MOVIE

I wish to start from an event in Moscow University about half a century ago. At that time, eminent professors mostly were not willing to give general lectures for newcomers and freshmen. Only few did it. Bigger auditoriums were provided for them, from a capacity of twenty to a hundred then to five hundred students. Eventually, students could not see and hear the lecturer.

One brilliant chap came up with an idea: to record the better lecturers on movie film and show the movies in dark rooms. At first, all the people were happy with the idea. The minister was happy due to salary saving because a movie operator earned much less than a professor. Students were happy because to sleep in a dark room or to kiss a girl in a dark room was much more convenient than in the light. But the idea was not implemented. Why not? Who objected? Wise professors. For every wise lecturer, it is absolutely necessary to see the attentive eyes of students and hear the questions of curious boys and girls. I underline *curious*. I'm sure you know the proverb *curiosity will kill the cat*. But I think: not the student. Curious students will change the world.

CURIOUS YOUNG PEOPLE

Let us recall some curious young people, who studied mechanical engineering or closely related disciplines. One was the son of a hydraulic engineer, who told his son about water wheels of ancient times and water turbines, where falling water produced much work. The power that could be produced was the product of the height of the dam and the water flow rate.

The curious son thought about heat engines. He understood that the height of water was similar to the height of mercury in a glass thermometer. It was more difficult to find the analogy for the water flow rate. He accepted the idea of a weightless fluid: caloric. The product of caloric flow rate and temperature difference was the power of a heat engine. A little later, he recognized that the efficiency of an ideal heat engine was the ratio of this power to the product of caloric flow rate and absolute temperature.

In his subsequent thesis, *Reflections On the Moving Force of Fire* in 1824, the only book published by him, classical thermodynamics was founded. He died from cholera in 1832 and all his things were incinerated, but luckily, his brother saved his notebook with lots of important thoughts.

The name of this man, best known in mechanical engineering, was Sadi Carnot. His eminent father, Lazar Carnot, was a well-known person in France in Napoleonic times, a minister during the notorious 100 days. The grandson of Sadi Carnot, also Sadi, was the President of France at the end of the nineteenth century.

Another curious boy studied mechanical engineering at Zurich Polytechnic Institute. But long before that, while still a teenager in his school years, in spite of a modest mark in physics, he thought about what a mirror reflected, when the mirror moved with a speed greater than the well-known speed of electromagnetic waves, the speed of light. If the mirror was faster, it should reflect the image of our grand-grandparents and old events, too. It was so fantastic and interesting! Twelve years later, in 1905, he had a job as a patent clerk in Bern, examining inventions, mostly in mechanical engineering. By that time, he understood that to overtake light is impossible and published a paper, *To Electrodynamics of Moving Bodies*. It was the foundation of relativistic mechanics. The name of this curious boy was Albert Einstein.

Why only boys? There are curious girls, too. One wonderful girl was a housemaid and then a teacher in a family and fell in love with the eldest son. They intended to get married, but the parents of the boy were against it and, in a sad mood, she left her native city and went to Paris. There, she entered the university and got married to a young professor. It was fantastically interesting to search for the physical reason of radiation. No one physical change affected that radiation. She separated a very small part from many tons of raw materials and discovered that an unknown element, radium, was responsible for the radiation. Her name was Maria Sklodovska, later Madame Curie. Her discovery paved the way for nuclear power, bombs, etc. That is why I suggest young girls should be allowed to get married in the university of their native city, without going to Paris.

A list of great mechanical engineers would include such names as:

James Watt — steam piston engine
Charles Parsons — steam turbine
Rudolf Diesel — piston engine with ignition from compression
Wright Brothers — aircraft in 1903
Sergey Koroliov — space rocket in 1957

S. KOROLIOV

I'm sure the first four names are familiar to you and your teachers. The fifth is much less known. The fate of this mechanical engineer was as tragic as it was brilliant; it deserves more details to be described.

Sergey Koroliov was born in 1906, graduated from Moscow High Technical School in Aircraft Mechanical Engineering in 1930, and in the daytime, worked in a design bureau. All his evenings and holidays were devoted to the study of rocket flight and to the construction and testing of liquid-fuel rockets.

His group was entitled Group of Research of Rocket Flight. Some sceptics called it Group of Engineers Working for Nothing.

In 1933, his rocket reached 600 m of altitude. In 1938, the Rocket Research Institute was established, based on this group. In the same year, due to Stalin's purge, the military marshal who founded the institute was executed. The institute was closed with a clear sentence for its staff: over 40 years of age: to death; less than 40 of age: to work in a lead mine. Koroliov was less than 40 and was sent to work in the lead mine, where nobody could survive more than one year. His mother asked the eminent pilot, Gromov (who flew over the North Pole to the U.S.A. from Moscow in 1937), for help. He asked the prime minister to do something and the work at the lead mine for Koroliov was replaced by work in custody in a design bureau. Many excellent engineers worked there compulsorily, living behind bars and without contact with their families and friends. This continued for 10 years. In 1948, Koroliov was released, commissioned in the uniform of colonel, and sent to look at the German Rocket Centre of Verner von Braun at Peenemünde in Germany. He was then given full governmental support to create a very big design bureau and a factory with powerful manufacturing capacity. The bureau designed big ballistic rockets, which in 1957, launched the first satellite, Sputnik, and in 1961, sent the first man into space (Gagarin).

Till his death in 1966 from an improper surgical operation, Sergey Koroliov was absolutely unknown in the world. His work had been top secret. In his time, Koroliov had orchestrated the work of about 100,000 people at his bureau and factory and at lots of other research institutes. He had the title chief constructor. The Nobel Prize for the first satellite was never awarded to Koroliov. I was privileged to have had a talk with him before his great successes.

ABOUT MYSELF

My fate is absolutely incomparable with the above-mentioned ones. In general, I was unsuccessful. A good beginning was followed by very small results. But my fate might be a useful negative example for education of modern curious boys and girls.

I was born in 1929 in Kharkov, Ukraine, a big city with a great industrial and scientific infrastructure. My first love was aviation. In 1951, I had been through Kharkov Aviation Institute where I trained as a mechanical engineer for aircraft construction. In the local aero club, I piloted a small plane. In total, my flight time was 120 hours, during 80 hours of which I piloted myself alone or with a passenger. I made all the figures of high pilotage.

My diploma project was a self-guided cruise missile, a small unmanned plane with infrared eyes that directed the missile to the hot spot in the sky created by the jets of the enemy's aircraft. The eyes themselves were beyond my scope: they were taken from German missiles.

After our diploma project examinations (known as defending ceremonies), two of my friends with similar projects and I were recommended to work for Koroliov by our professor, a close friend of Koroliov's from his period in custody. Koroliov complimented us on our projects, but turned us down. He told us, "I am not so powerful as to be able to call off your commissioning by the Ministry." This was in the summer of 1951, long before the first satellite and the Gagarin flight. So I was obliged, according to my commissioning, to go to the city of Taganrog at the Azov Sea to a mill or factory, which was manufacturing big anti-submarine flying boats, where I worked as a technologist.

As is true for all first loves, my love of aviation ended, but has remained in my memory forever. Very occasionally, I was invited to work in a big electromechanical mill, producing electrical machines of medium and big sizes. In a design bureau of the mill, there were many excellent engineers who were knowledgeable about magnetic fluxes and electrical currents, but were unaware of how to perform calculations relating to heat and airflows. They needed a mechanical engineer for this work. The journal bearings of big machines, where shafts were rotating in oil, were within my scope as well.

I devised an algorithm to calculate heat flow in asynchronous motors of the closed type. Today, it would be called a computer program, but at that time, our computer was a logarithmic lineal (a slide rule) only.

I remember a problem with the thrust bearing of a big machine with a vertical rotor, which was driving a water pump for irrigation. As usual, it was of the Mitchell type with six tilting pads.

They were to bear the thrust force. The usual cylindrical journal bearings carried radial forces. I decided to replace the two bearings with one spherical bearing in an oil bath, which could catch the axial and radial forces together. I prepared the drawings. My boss, who was the head of the design group and an experienced engineer, was flatly against this innovation as it had not been proven beforehand.

The work was his responsibility and he was quite justified in his doubts. But he did not know any foreign languages (as was the case for the majority of Russian engineers at that time) and I showed him a picture from a German magazine with something spherical in it and assured him that it was a bearing. Allowing a spherical bearing to be used was a big risk to my boss, but he took that risk because of my encouragement. Showing him a magazine picture that was not a bearing was a bluff on my part, but I was too young and too curious. Many machines were built with my spherical bearings. Subsequently, I saw some of them working on site.

In 1959, I was commissioned to go to the city of Norilsk, at 69 degrees latitude, in the tundra. It was only three years after the closing of the most terrible concentration camps of the gulag system. Many traces remained, but the city itself was surprisingly beautiful, some parts were like St. Petersburg.

There had been damage to a big direct-current generator, which supplied an electrolytic bath for producing pure nickel. The machine was overheating. I installed a new fan beneath its collector and put it in order. That was in March. Our car from the city to the airport was driving in a canyon between ten-meter-height walls of dense snow. If there had been a snowfall, the car might have been fully covered and stopped. Luckily, this was not the case.

MAGNETOHYDRODYNAMIC (MHD) GENERATORS

Most important for my career was an idea I had in 1955, when nightly I tested a closed-type asynchronous motor, measuring the temperature of its squirrel-gage type rotor. I had seen temperatures of 500°C, where the melting point was 710°C. Let it melt … what would happen? In liquid metal, the conductivity is the same as in solid metal, which meant the currents and forces would be the same, but liquid could flow!

This meant that an electromagnetic pump was possible. Soon, I had found a patent of 1929 by L. Szilard and A. Einstein concerning such pumps. Then, I learned about the use of such pumps in some nuclear reactors (fast breeders with sodium cooling).

If a pump (analogous to a motor) was possible, a generator (analogous to a turbine) was possible, too. Had we been able to have enough conductivity in a gas (by converting it into a plasma state), such a generator might replace a gas turbine. The weak point of the latter is the vulnerability of its blades in high-temperature gas flow. But for a magnetic field, a high-temperature flow is not detrimental. In the same year,

1955, was published the crucial observation by A. Kantrovitz and E. Resler on the extremely high electrical conductivity of inert gases, like argon, in a shock wave. They explained it was caused by the evaporation of the shock tube's metal walls and the low ionization potential of metal vapors.

In a handbook, I had found some metals that had the smallest ionization potentials (cesium and potassium) and had evaluated the conductivity of mixtures like argon or neon (which had the smallest collision cross-section with electrons) with small additions of cesium or potassium. The conductivity was enough for the strong interaction of a magnetic field with such flowing gases. I got a Russian Author's Certificate (analogous to a western patent), dated 1959, for the best working substance for magnetogasdynamic machines and some alternating current magnetohydrodynamic (MHD) generators. In the same year, a paper by A. Kantrovitz appeared concerning a test of a direct current MHD generator in the company AVCO-Everette in the United States. Luckily, in the same year, a research institute was established at the Kharkov mill. The MHD laboratory started there in 1959 and I headed it over the next 10 years. The laboratory still exists today and is active in the area of applied magnetohydrodynamics.

Very soon in this laboratory, we had made and installed our test rigs with a 5-ton magnetic system and a 1000 kW plasmatron to create plasma flows at 4000 K. In 1961, we demonstrated a small bulb energized by plasma flow in a magnetic field. Over some years, we developed the design of a utility scale MHD generator with high voltage generation. In Moscow, a very rich and powerful Institute of High Temperature had built a full pilot power plant with an MHD generator of 20 MW. In the media, the plant was described as a "Power Plant of the year 2000." This prototype magnetohydrodynamic generator could operate for no more than half an hour before damage to the flow channel became excessive. Unfortunately for us, the engineering problems of MHD channel construction turned out to be insurmountable. Nobody in the world could create a stable channel for plasma flow with electrical arcs near electrodes. All such research around the world, including that in the U.S.A. and in the U.K., was cancelled.

Another conducting fluid, molten metal, was less demanding. When S. Koroliov decided to start a project for a manned Martian mission in 1961, I got an order to build a liquid metal MHD generator of alternating current to convert nuclear energy of a small fast breeder reactor into electricity for the electrical thrusters. The working substance was to be a mixture of sodium and potassium at a temperature of 600°C, with a design life of some years. We in the MHD laboratory (I. Tolmach, L. Dronnik, and I) designed, built, and tested in a vacuum chamber such a 5-kW generator operating at 600°C, working for some hours. The chief engineer of the Baikonur Rocket Cosmodrome took this machine in his hands and assured us that it was light enough. However, the Martian mission was postponed for a long time. In 1966, S. Koroliov passed away. In 1991, Communism in Russia passed away and our really fantastic machine was used for spin-off technologies.

Heat Pumps

In 1970, I moved to Moscow to work for a leading institute in energy engineering, named after Krjijanovski. There, I got a full doctorate degree. I had to analyze new energy technology, including renewable and thermonuclear options. In 1974, I was invited to the Institute for Industrial Energetics to head the Heat Pump Laboratory.

There were millions of heat pumps in operation in the world at that time, especially in Japan, but not one in the Soviet Union. I understood that for economical attractiveness, a suitable low-grade heat source was of primary importance. We had found such a source with a temperature in January of 15°C. This was a clean water stream from Moscow's Aeration stations.

We intended to replace some old inefficient, air-polluting boiler houses with heat pumps, but the capacity of heat pumps would have to be increased from 1 to 3 kW to 10 to 30 MW. We organized an international workshop in April 1978, where there were ten attendees from major European countries. There, I delivered our detailed calculations concerning such heat pump stations (HPSs) using water flows as the low-grade heat source. Some years afterward, such HPSs were widely built in Sweden and gave huge economic benefits. They replaced oil-fired heating with heating provided from cheap hydropower electricity with three times more thermal kilowatts than the consumed electrical kilowatts. In the Soviet Union, we were able to build two HPSs of 3 MW each, both in Krimea, near Yalta, using water from the Black Sea, which had a temperature of 10°C in winter.

Exergonomics

I had realized the benefit of the exergy concept for heat pumps and started to examine the exergy concept for power systems. In 1989, I was granted a professorship in Energy Systems and Complexes by the State Attestation Committee. At that time, I worked in the Institute for Energy Research of the Russian Academy of Sciences. In such institutes, there are no students; there is only research. I asked some authorities what I should do as a professor. They told me, "You should lecture to university lecturers." I developed a course "Physical Background of Energy Engineering," where the exergy current vector played a major role, and started to lecture at a big seminar for university lecturers. At the first lecture, I saw about 100 listeners; at the second 30; at the third 10.

I asked an experienced colleague what was the reason for such a decline. He explained: "They prefer to teach. It is much easier than to learn. When learning, you need to hear and understand; when teaching, you only need to speak."

While I was visiting Duke University in the U.S.A., I found the book by Frederick Soddy entitled *Wealth, Virtual Wealth, and Debts*, of 1926, and learned his life story. He tried to replace money in the evaluation of goods and services by a physical quantity: content of energy. I thought, "Energy is conserved in insulated flow and cannot be destroyed; exergy is better." But in Soddy's time, the exergy concept was not understood. In 1994, I published a book *Energy and Exergy Currents* (introduction in Exergonomics) in the U.S.A. I intended to make a mirror image of economic calculations of energy projects with the replacement of monetary value by exergy. For decision makers, the most interesting approach might be optimization in the three-dimensional frame exergy-money-pollution (Pareto optimization), which was published in 2000.

ZERO EMISSIONS TECHNOLOGY

Over the last decades, I have tried to select the best way forward for energy supply. I was aware of the weak point of nuclear power long before our terroristic era. The Tchernobyl disaster confirmed my fear.

Having looked at all renewables like solar, wind, and ocean, I understood that they are good in some places but, even taken together, could not meet the world demand over the next hundred years. The use of fossil fuel continues to be the only realistic option. Evidently, the forecast of the Club of Rome in 1972 that worldwide oil extraction would peak in 1995 and then fall off was wrong. All current forecasts suggest a linear increase in the use of fossil fuels, especially for natural gas. The only real problem of fuel use is the release (emission) of carbon dioxide, causing global warming.

After many years of controversies, global warming is now an evident reality as evidenced by floods and droughts. After understanding what the problem is, the next question should be what is the solution? I decided to spend the rest of my life searching for the answer. In 1991, I was privileged to give a lecture at the World Clean Energy Conference in Geneva, *The Thermodynamics of Fuel-Fired Power Plants Without Exhaust Gases*. I discussed some new schemes with air separation at the inlet, combustion of gaseous fuel in a mixture of oxygen and steam, triple-stage steam turbine expansion, water recirculation, and carbon dioxide separation and sequestering. Five years later, persons who had been part of a group of U.S.-based German rocket engineers patented such a scheme. They set up Clean Energy Systems, Inc. in Sacramento, U.S.A. Now, they are preparing to demonstrate a small unit of this type.

A crucial problem of such power units, which belong to zero emissions technology, is the production of oxygen. I have performed calculations for many such power cycles with three groups of co-workers. A significant weakness of the concept is the requirement for an air separation unit of the cryogenic type. This method of air separation is a mature technology, but is rather expensive in energy and cost.

Quite recently, I and my colleagues J. Gorski, B. Smyth, and J. ten Elshof developed a new possibility: the use of ion transport membranes for oxygen production. The new cycle is entitled ZEITMOP. A paper on this was published at the ECOS 2002 Conference in Berlin and at the Carbon Sequestration Conference, in Washington, U.S.A. This cycle is very promising: its further elaboration and associated experiments on combustion inside a membrane tube would seem to be a suitable subject for research at an institute such as the DIT.

EMIGRATION

Since 1990, I have often been invited to lecture in the U.S.A. and Europe (MIT, Tennessee Tech, Miami University, Florida University, Duke University, Vesteros, Trondheim, Nancy, Utrecht University). I have spoken mainly on exergonomics and zero emissions power.

In 1994, my institute in Moscow was in a critical state due to the lack of money to pay salaries. A rumor about me arose: that I was swimming in dollars and that with earnings from abroad, I might live without salary. I was forced to apply for a vacation. I could not wait further and emigrated to Germany with my wife in April 1995. In Germany, I feel quite comfortable with a small pension. Sometimes, I work as a visiting professor (Liege University, 1997–1998).

I will conclude now. From my modest practice of education, I have learned: a student is not a jug, which should be filled, but a torch, which should be ignited. My dear colleagues, future mechanical engineers, or future engineers of other disciplines, I have done my best to ignite you. Don't be too wet!

11 Concluding Remarks

After considering many zero emission power cycles, one may ask whether it is possible to reach the ultimate goal, to create in some future the zero emissions city, using ZEPP as the main element. No one stack or exhaust pipe in the atmosphere should exist in such a city.

Continuing the line, "to ban emissions," we have to keep in mind all four sources of emissions: power generation, fleet vehicles, heating of dwellings, and industry (metal, cement, refinery). The last one needs to introduce some special changes, in particular technologies, which is beyond the scope of this book. We wish to indicate what could be done in cities with the first three sources, when a city is free of nearby heavy industry.

The political sector will be in favor of ZEPP (Carbon Capture and Sequestration CCS), which is now being urged by high ranking bodies, as in the U.K. For example:

The Environmental Audit Committee, an influential parliamentary body, has published a report that strongly urges the UK government to set a deadline for all coal fired power stations be equipped with carbon capture and storage technology or face closure... A group of leading British scientists ... have urged the U.K. government, in an open letter, to deploy CCS technology as quickly as possible, warning that unless it acts with urgency, it risk losing a world lead in carbon capture technology to other countries, such as Canada and Germany, (PEi 6/2008, p.5.)

The most important element of the system is ZEPP, described in Chapters 2-6. The ZEMPES is described in Chapter 7. The electrical car is now presented in many papers. One of the best examples is the Mercedes-A-klass with ZEBRA batteries; it has a range of 240 km and achieves 50 km/hours in 4 sec at the start.

For the heating of dwellings without emissions there exist two well-known possibilities: cogeneration of ZEPP (with hot water pipe to dwelling) and heat pumps, using low grade heat from air, sea water or ground. A good example is the heat pump station of 180 MW_{th} in Stockholm, which uses low grade heat from the Baltic Sea with a temperature of 2°C in the winter.

In some distant future, when not only fuel deficiency, but also oxygen depletion in the atmosphere, could create problems, the oxy-fuel ZEPP might be replaced by SOFT cycle.

This cycle converts solar energy by photosynthesis in macroalgae with combustion of produced organic matter in oxy-fuel ZEPP (Chapter 8). It is free of any natural restrictions as it uses neither fossil, nuclear fuel, oxygen, nor much land. The SOFT cycle really belongs to the noble family of renewable energy sources.

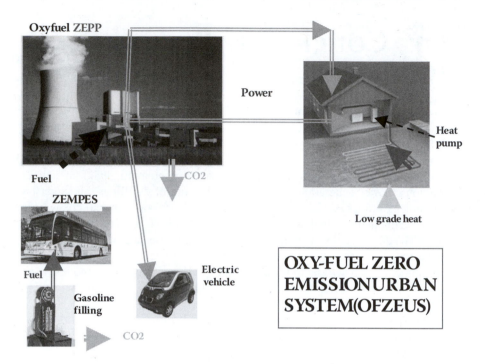

FIGURE 11.1 Schematics of the energy supply CCS system of a zero emissions city.

A most important recent event is the start of 09.09.2008, the world's first coal-fired oxy-fuel ZEPP made of 30 MW by Vattenfall in Germany (Schwarze Pumpe). It will encourage many bigger projects.

The schematics of the zero emissions city is presented in the final figure of the book, Figure 11.1. It presents the ultimate goal of the creation of zero emissions power plants as unavoidable elements of the future sustainable development of mankind.

There have been some very exciting events occurring recently in Europe. The European Commission Industry scientists have united to form the European Technology Platform for Zero Emissions Fossil Fuel Power Plants (ETP-ZEP) with the goal to enable fossil fuel power plants to have zero CO_2 emissions by 2020. The world's first coal-fired pilot plant of 30 MW by Vattenfall, which was mentioned in the book, was successfully commissioned in September 2008. The great coal-fired power plant Niederaussem by RWE got started with its first use of seaweed-algae to convert CO_2 into biomass by photosynthesis, with an attempt to form a closed cycle zero emissions process, as mentioned in Chapter 8. With regard to renewables, Germany is a good example to the rest of the world when it comes to ZEPP development.

Index

A

Abiogenic theory, 10, 15
  drilling experiment, 12
Adiabatic compressor, 113
Adiabatic engines, 148
Advanced technology partial zero emissions
    vehicle (ATPZEV), 142
Advanced Zero Emission Power (AZEP)
  cycle, 35, 44, 98
  development, 99
  ITM reactor design, 99, 100
  project, 148, 149
  reactor, 99, 100, 101
  technical data, 101
Aero-gas turbine cooling, 127
AFDW. *see* Ash free dry weight (AFDW)
Afterburner, 151, 152, 153, 164
  AK, 152, 157
  thermal power, 169
  VK, 152, 164
Air
  adsorptive separation process, 166
  compressed, 131
Air Products, 96, 143, 153
  and chemicals, 171
Air separating unit (ASU), 178
  external cryogenic, 85
  ITM, 107
Aker Company, Norway, 27
Aker Maritime, 33
Albany Research Center, 15
Alexejev, V., 54
Algae
  ash content, 185
  calorific content, 185
  cultivation and use, 179
  heating values, 185
  photosynthesis, 194
  photosynthesis conversion, 179
  photovoltaics, 188
  pond, 188
  zero emission oxy-fuel combustion solar
      energy conversion, 179
Allam, R., 96
ALSTOM Power Inc., 33, 45, 149
Aluminum conductor, 222
Ambient air, 163

AMR. *see* Automotive membrane reactor (AMR)
AMSTWEG. *see* Automotive membrane
    steam turbine without exhaust gases
    (AMSTWEG)
Argon, 267
Argonne Lab, 22
Arons, G., 219
Arrhenius, Svante, 4, 258
Artificial air, 144, 148, 163, 180, 260
Ash content of algae, 185
Ash free dry weight (AFDW), 183
Aspen Plus, 126, 130
  ZEITMOP-combined simulation, 131
ASU. *see* Air separating unit (ASU)
Asynchronous motor, 266
Atmosphere, 4
ATPZEV. *see* Advanced technology partial zero
    emissions vehicle (ATPZEV)
Automotive membrane reactor (AMR), 143, 158
  heat flows, 146
Automotive membrane steam turbine without
    exhaust gases (AMSTWEG)
  cycle, 55, 118
  nondimensional flow rates, 55
Auxiliary Brayton cycle, 163
Auxiliary cycle diagram, 165
Available energy, 206
AVCO-Everette, 267
  laboratory, 123
Axial-main compressor, 123
AZEP. *see* Advanced Zero Emission Power
    (AZEP)

B

Baikonur Rocket Cosmodrome, 267
Barentz Sea, 16
Barium oxide, strontium, cobalt, ferrum (BSCF)
    ceramic, 101
  operating conditions, 103
  tubes, 101, 102
Beichel, Rudolf, 261
Biogenic theory, 10, 12
Biomass
  carbon dioxide, 62
  combustion, 177–178, 180
  defined, 177
  fired zero emissions cycle, 104, 105